THE ODD QUANTUM

SAM TREIMAN

The Odd Quantum

Princeton University Press, Princeton, New Jersey

Published by Princeton University Press, 41 William Street,
Princeton, New Jersey 08540
In the United Kingdom: Princeton University Press,
Chichester, West Sussex

Library of Congress Cataloging-in-Publication Data

Treiman, Sam B.
The odd quantum / Sam Treiman.
p. cm.
Includes index.
ISBN: 0-691-00926-0 (cl : alk. paper)
1. Quantum theory. I. Title.
QC174.12.T73 1999
530.12—dc21 99-24123

This book has been composed in Palatino

The paper used in this publication meets
the minimum requirements of
ANSI/NISO Z39.48-1992 (R 1997)
(*Permanence of Paper*)

http://pup.princeton.edu

Printed in the United States of America

10 8 6 4 2 1 3 5 7 9

CONTENTS

PREFACE

This book suggested itself after I had conducted a one-time, one-semester freshman seminar at Princeton University. The seminar program, open only to first-year students, offers a wide range of special topics, many of them quite ambitious. Student participation is voluntary and selective; class sizes are small. The seminar in question was entitled "From Atoms to Quarks, Along the Quantum Trail." I had anticipated and the students later confirmed that the material was rather demanding. But they were eager, open, and numerate. Most had taken earlier plunges of various depths into the popular literature on relativity, cosmology, the atom, nuclear and particle physics, and so on; and some had gotten whiffs of these subjects in high-school courses. They wanted to know more. It seemed likely that several of the students would later on, in the sophomore year, elect to major in one or another of the natural sciences or engineering. Others were headed in other directions, in the social sciences or humanities. What they had in common was a curiosity about atoms and electrons and neutrinos and quarks and quantum mechanics and relativity, and all that.

For many of the topics covered in the seminar there were excellent readings to be recommended, in books that offer mainly descriptive, not-too-mathematical accounts of the development of the atomic hypothesis in the nineteenth century; the subsequent discovery of the nucleus and its components; the later flood of subnuclear particles of various kinds; the modern quark picture; and so on. However, in order to dig beneath the qualitative picture and provide a deeper understanding, I wanted to devote some time to the underlying theoretical framework, to an introduction to quantum mechanical concepts and practices. There is of course no shortage of quantum mechanics textbooks for undergraduate majors, graduate students, and professionals in various branches of science and technology. In the other direction, there are many wonderful books in which the exposition of quantum mechanics relies chiefly on qualitative descriptions, analogies, metaphors, allusions, and the like. Many employ imaginative graphics, include interesting biographical sketches of the founders, and employ other devices to capture the reader's interest.

What I could not so readily find are books that lie in between; treatments, that is, that are sufficiently probing and mathematical to convey something of the actual substance, methods, and oddities of quantum mechanics, yet not overly technical or professional. The present slim volume has these in-between objectives as its goal. It is aimed at a wide audience of the curious—scientists in non-quantum-mechanical disciplines as well as nonscientists—at any rate those in either class who are not put off by equations and technical particulars. It certainly goes beyond those freshman, but they might have dipped into it. I will be pleased if the book is received as a series of related, short essays. A word about the mathematics: it is here to give explicit form to concepts that are often best grasped through their concrete expression in equations and in the interpretations that go with those equations. For example, it is one thing to assert vaguely that quantum mechanics deals with probabilities, another to embody this in a definite mathematical object, a wave function whose evolution in time is governed by a definite equation and whose information content is spelled out in terms that are sometimes, of necessity, mathematical. The reader is not much asked to actually solve any equations other than easy ones, but the reader is invited from time to time—optionally—to confirm a solution that is provided gratis.

Quantum mechanics is the main theme of the book; but I could not resist the temptation to indulge in brief reviews of classical mechanics and electromagnetism, special relativity theory, particle physics, and other topics.

I am grateful to Joan Treiman for her encouraging words, and for her forbearance.

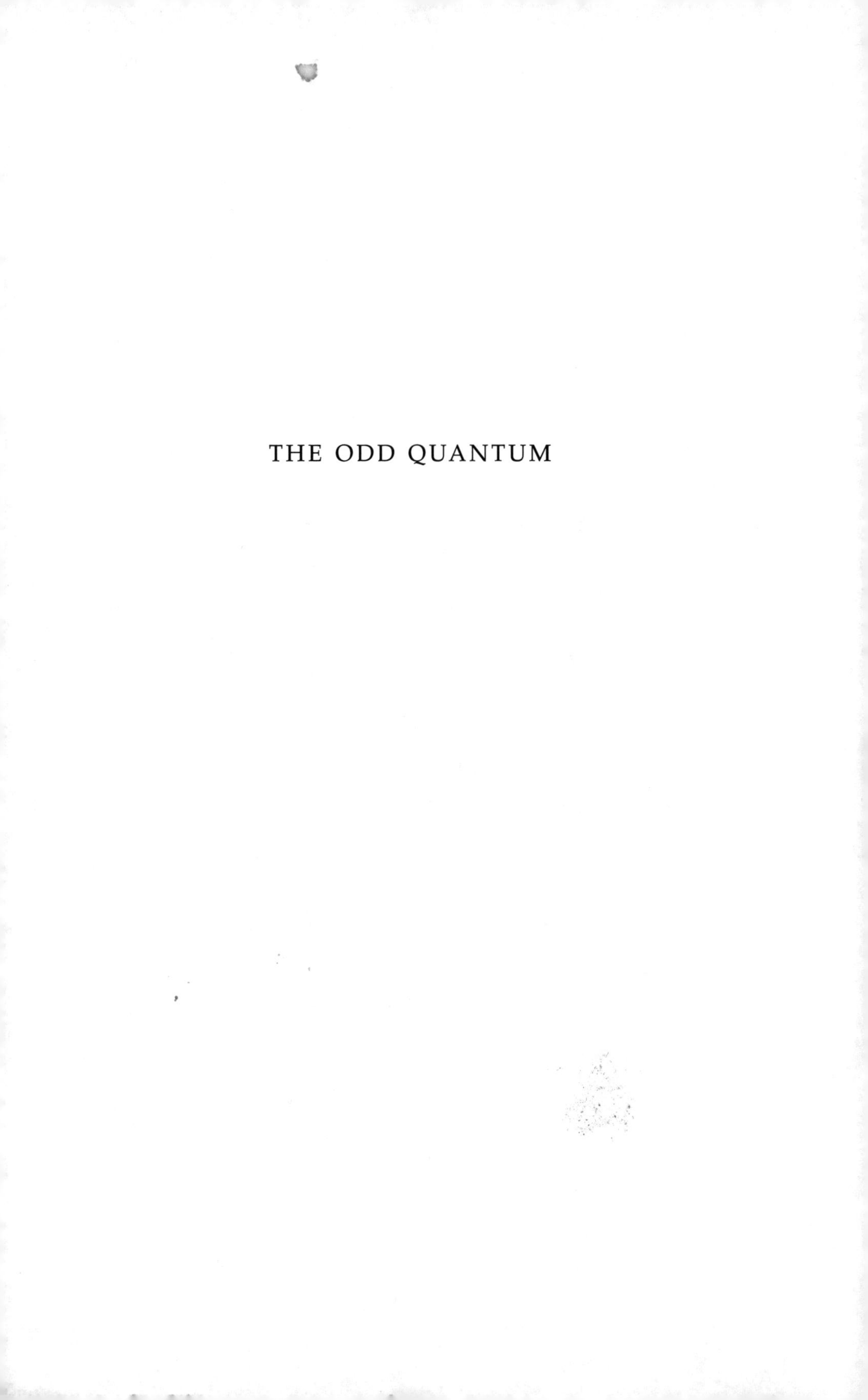

THE ODD QUANTUM

CHAPTER ONE

Introduction

In the physics section of the University of Chicago catalog for 1898–99, one reads the following:

> While it is never safe to affirm that the future of the Physical Sciences has no marvels in store even more astonishing than those of the past, it seems probable that most of the grand underlying principles have been firmly established and that further advances are to be sought chiefly in the rigorous application of these principles to all the phenomena which come under our notice. . . . An eminent physicist has remarked that the future truths of Physical Science are to be looked for in the sixth place of decimals.

This catalog description was almost surely written by Albert A. Michelson, who was then head of the physics department and who had spoken very nearly the same words in a convocation address in 1894. The eminent gentleman whom he quotes may well have been Lord Kelvin. That 1894 talk proved to be well timed for contradiction. In quick succession, beginning soon afterward, there came the discovery of X-rays, radioactivity, the electron, special relativity, and the beginnings of quantum mechanics—all of this within a decade centered around the turn of the century. Indeed, it was Michelson himself, working together with E. W. Morley, who in 1881 had carried out the crucial experiment that was later recognized as a foundation stone of special relativity. Both Michelson and Kelvin received Nobel Prize awards in the early years of the twentieth century.

In short, all the grand underlying principles had *not* been firmly established by the end of the nineteenth century. This cautionary tale should not be told with any sense of mockery. Those distinguished scientists—and there were others who spoke along the same lines—were looking back on a century of extraordinary accomplishment, an epoch that had carried the physical sciences to a state of high development by the late years of the century. The wavelike character of light had been demonstrated; the laws of electricity and magnetism were discovered and placed together in a unified framework; light was shown to be the manifestation of electric and magnetic field oscillations; the atomic hypothesis had increasingly taken hold as the century moved on; the laws of thermodynamics were successfully formulated and—for atomists—grounded in the dynamics of molecular motion; and more. To be sure, although the gravitational and electromagnetic force laws seemed well understood, it remained yet to learn whether other kinds of forces come into play at the atomic level. That is, there was work yet to be done, and not just at the sixth place of decimals. But a clocklike Newtonian framework seemed assured. In this *classical* picture of the physical world, space and time are absolute; and every bit of ponderable matter is at every instant at some definite place, moving with some definite velocity along some definite path, all governed by the relevant force laws according to Newton.

This classical outlook in fact continues to provide an excellent description of the physical world under conditions where velocities are small compared to the speed of light and relevant dimensions large compared to the size of atoms. But our deeper conceptions of space-time have been transformed by relativity; and of objective reality, by quantum mechanics. Both run counter to everyday experience, to our common sense of the world. This is especially so for quantum mechanics, which is the focus of the present book.

Overview

Before we embark on our journey, it may be good in advance to sketch out very roughly some of the contrasts that will be encountered between the classical and quantum modes. For the most part here, we will be considering a system of point particles moving under the influence of interparticle and perhaps external force fields characterized by a potential energy function.

Quantization

Classically, a particle might be anywhere a priori; and it might have any momentum (momentum = mass × velocity). Correspondingly, its angular momentum—a quantity defined in terms of position and momentum—might a priori have any value. So too the particle's energy, kinetic plus potential, might have any value above some minimum determined by the potential. Quantum mechanically, however, angular momentum can take on only certain discrete values. It is *quantized*. Energy is sometimes quantized too, depending on details of the force field. It is this classically inexplicable discretization that provides the adjective "quantum" in quantum mechanics.

Probability

A much sharper and more profound contrast with classical mechanics has to do with the probabilistic character of quantum mechanics. For a classical system of particles, the state of affairs is completely specified at any instant by the position and momentum variables of all the particles. The data on positions and momenta at any instant constitute what we may call the *state* of the system at that instant. It tells all that can be known dynamically about the system. Other quantities of interest, such as energy, angular momentum, and so on, are defined in terms of the position and momentum variables. Classical mechanics is deterministic in the sense that future states of the system are fully and uniquely determined if the state is specified at some initial instant. The present determines the future. Of course, in practical situations the initial data will inevitably be compromised to some greater or lesser extent by measurement uncertainties. Depending on the system under consideration, the future may or may not be sensitive to this uncertainty. But there is no limit *in principle* to the accuracy that can be imagined. There is no bar in principle, that is, to precise knowledge of the position and momentum of each particle, and therefore no bar to anticipating future developments. When wearing our classical, commonsense hats, we do not doubt that every bit of matter is at every instant at some definite place, moving with some definite momentum, whether or not we are there to observe.

The notion of state also arises in quantum mechanics. Here again the *state* of a system connotes *all that can possibly be known about the system at any instant*. Also, just as in the classical case, the system develops deterministically in that future states are fully determined if the state at some initial instant is given. In this sense, here too the present determines the future. But there is a very profound difference. A quantum state does not precisely specify particle positions and momenta, it only specifies probabilities. Quantum mechanics, that is, is probabilistic! For example, there are states for which the probability distribution of a particle's position is sharply localized, so that the position may be said to be almost definite (at the instant in question). But there are other states for which the probability distribution is broad, so that upon measurement the particle might be found almost anywhere. And there are infinitely many possibilities in between. So too for momentum: for some states the momentum is almost definite, for others it is broad, and there are infinitely many possibilities in between.

This probabilistic description obtains not because we have imperfect information about the state of the system, but is intrinsic. Moreover, the rules of probability composition have some very peculiar features. We will, of course, go into these things more fully later on, but it is important already at this early stage to emphasize a point that may be illustrated with the following example. Suppose one sets up detectors at various locations to determine the position of a particle known (somehow) to be in a certain quantum state at a certain instant. If a particular detector clicks, we will have learned that the particle was in the volume occupied by that detector at the instant in question. That is, there *will* be a definite finding of location. But if the experiment is repeated over and over, always with the particle arranged to be in exactly the same state, there will be a spread of outcomes. On different runs different detectors will click. Full knowledge of the quantum state does not allow one to predict the outcome event by event, only the probability distribution.

The Uncertainty Principle

It is the case that any state that has a very localized probability distribution for position measurements will inevitably have a broad distribution for momentum measurements, and vice versa. There is a limit to how well one can jointly localize both position and momentum. So too for certain other pairs of *observables* (as measurable quantities are called). This is enshrined in the celebrated Heisenberg uncertainty principle. That principle is not some add-on to quantum mechanics; it is a technical consequence that flows from the structure of quantum mechanics. As must of course be the case, for the macroscopic objects of everyday life the Heisenberg limit is not at all a practical restriction. We can, for example, know both the position and momentum of a moving jelly bean quite accurately enough for all everyday purposes. However, at the atomic level the uncertainty principle comes fully into play.

Identical Particles

In the macroscopic world, we never encounter two or more objects that are strictly identical in every possible respect: mass, composition, shape, color, electric charge, and so on. But even if we did—and we do at the microscopic level, where, for example, one electron is exactly the same as another—this would pose no conceptual problem for classical science. One can in principle keep separate track of the objects by, so to speak, pointing: object 1 is the one that's at this place, object 2 is the other one over there, and so on. For quantum mechanics this approach has its limits. It is not possible to keep track in this way since locations are a probabilistic matter. Rather, there is a distinctly quantum mechanical approach to dealing with identity, one without classical analog. The implications are sometimes quite unintuitive, and they are profound. What is most remarkable is that all the known particles indeed come in strictly identical copies—all electrons are the same, all protons the same, and so on. Quantum field theory provides the only natural explanation for this striking fact of identity.

Radioactivity

This term refers to processes in which an atom *spontaneously* emits one or more particles: an alpha particle (helium nucleus) in the case of one class of processes, α decay; an electron (plus neutrino as we now know) in another class, β decay; an energetic photon in yet another class, γ decay. In α and β radioactivity, the parent atom is transmuted in the process into a daughter atom of a different chemical species. There is no such transmutation in γ radioactivity. One speaks of any of these spontaneous events as a *decay* process. In the case of α and β radioactivity there really is decay, the disappearance of the parent atom and its replacement by an atom of a different ilk. In γ radioactivity the atom does not change its chemical species membership; but as we will see later, it does undergo a change from one energy level to another. In that sense, here too there is decay—of the occupancy of the initial energy level.

Not all atomic species are radioactive, but many are. When radioactivity was first discovered around the end of the nineteenth century, there was great wonder and bafflement. Many questions were raised, among them the question: where in the atom (*if* in the atom) do the ejected particles come from? This was clarified only after Rutherford formulated his famous model of the atom, picturing it as a swarm of electrons orbiting around a positively charged nucleus that is very tiny but that nevertheless carries most of the mass of the atom. With that, it soon became clear that radioactivity is a *nuclear* phenomenon. Two other questions among many remained, and they were especially puzzling: (1) The emitted particles typically carry a lot of energy. Where does that energy come from? (2) How does the nucleus decide when to decay? As to the first of these questions, the answer was already available in Einstein's 1905 formula $E = mc^2$; but it took a while before this sank in conceptually and before sufficiently accurate mass measurements of parent and daughter nuclei could be made to test the concept.

The deeper question (2) had to await the interpretative apparatus of quantum mechanics. If you take a collection of identical atoms of some radioactive species, you will find that the atoms do not all decay at some one characteristic instant but, rather, at various times—randomly. If the emissions are being detected by a counter, you may hear individual clicks as one or another atom decides to decay. As time goes by there will of course be fewer and fewer surviving parent atoms. As it turns out, the population of survivors decreases with time in an essentially exponential fashion, the average time (or, briefly, the *lifetime*) being characteristic of the particular species under consideration. On the classical outlook, the problem is this. The atoms of the given species are presumed to be identical. If they are governed by the clockwork regularity of classical science, why don't they all decay at the same instant, whatever may be the mechanism that causes radioactive decay?

The quantum mechanical answer is that the world is a probabilistic place. An ensemble of identical atoms starting in identical conditions will distribute their decays in a probabilistic way over time. One cannot predict what will happen event by event, atom by atom. What *can* be deduced quite generally is the exponential character of the decay curve. But the mean lifetime varies from species to species and depends sensitively on details of the underlying quantum dynamics. It should be said here that the traditional classes of nuclear instability, α, β, and γ, are only three among a much wider range of decay processes that occur in nature, including hordes of reactions involving subnuclear particles: pi meson decay, muon decay, and so on. The average lifetimes vary over an enormous range, from roughly 10^{-24} seconds for certain subnuclear particles to billions of years and more for certain α emitters (among these, U^{238}, whose half-life happens to be about the same as the age of the earth).

Tunneling

The probabilistic structure of quantum mechanics incorporates the possibility that a particle can be found in locations that are absolutely forbidden to it classically. For example, it can happen classically that there is an energy barrier that separates one region of space from another, so that particles below some energy threshold cannot penetrate the barrier and thus cannot move from one region to the other (it may take more energy than you've got to climb the hill that intervenes between where you are and where you want to go). Quantum mechanically, there is a finite probability that such strange things can happen. Particles can be found in, and can *tunnel* through, classically forbidden regions.

Antimatter

In attempting to find a relativistic generalization of Schroedinger's quantum equation for the electron, P. A. M. Dirac devised a theory that was spectacularly successful in its application to the hydrogen atom but that carried with it some seemingly bizarre baggage: among other things, negative energy states for the free electron. When properly reinterpreted this transformed itself into the prediction of a new particle having the same mass as the electron but opposite (that is, positive) charge. The antielectron, or *positron* as one calls it, was soon discovered experimentally. The situation has since become generalized. Relativistic quantum theory predicts that particles having electric charge must come in pairs with opposite charges but identical masses (and identical lifetimes if unstable). One member of the pair is called the particle, the other the antiparticle. Which is called by which name is a matter of history and convenience. It turns out that there are other kinds of "charge" in addition to electric charge; for example, so-called baryon number charge. The necessity of particle-antiparticle pairs obtains for charges of any kind. Thus, not only is there an antiproton to the proton, there is an antineutron to the neutron. The neutron is electrically neutral but it has baryon number charge. On the other hand, the photon and π^0 meson among others do not have antiparticles; or as one says, each is its own antiparticle.

Creationism, Destructionism

Our notion of what it means to say that something is made of other things has undergone a revolutionary transformation in this century. When you take a clock apart you find gears, springs, levers, and so on (or maybe a quartz crystal and battery). You say the clock is made of these parts. If you take apart the parts in finer and finer detail, you eventually get to atoms. If you take apart atoms, there are electrons and nuclei of various sorts. Going on, you find that the nuclei are made of protons and neutrons, and then that these are made of quarks and gluons. At the microscopic level, incidentally, taking apart means zapping the target with a projectile and looking at the pieces that emerge. In earlier years the surprise may have been that deconstruction did not stop at the atom. Still, the ancient notion could persist that, eventually, one comes to the immutable ingredients of the world, building blocks that can arrange and rearrange themselves in various combinations but that are themselves eternal and indestructible.

Thus, for example, the nuclear reaction $d + t \rightarrow \text{He} + n$ can be pictured as a mere rearrangement of the neutron (n) and proton (p) ingredients of the deuterium (d) and tritium (t) nuclei, the ingredients reemerging as the helium nucleus (He) with one neutron left over. The particle reaction (i) $\pi + p \rightarrow \Lambda + K$ might be taken to indicate that the particles involved here—pion, proton, lambda particle, kaon—are made of tinier things, perhaps quarks, that are similarly rearranging themselves. But if so, what does one make of the reaction (ii) $\pi + p \rightarrow \Lambda + K + \pi$, in which an extra pion appears on the right? Haven't the quarks already been conceptually "used up" to account for reaction (i), so that there are no ingredients left over to explain reaction (ii)? And what does one make of the reaction $p + p \rightarrow p + p + \pi^0$? No amount of rearrangement can explain how it is that the final system contains the same objects as the initial system *plus* something else. There is no getting around it, the π^0 is simply created here de novo; or at any rate its ingredients are. In short, down at the subnuclear level one is simply forced to acknowledge that particles can be created and destroyed!

This creation and destruction of matter is not something of everyday experience. It is a phenomenon that comes into play at high-energy particle accelerators, in the collisions induced by cosmic rays (high-energy particles that rain on the earth from outer space), in the stars and wider cosmos, and in certain radioactive decay processes. The transactions underlying most of science, technology, and everyday life have mostly to do with the "mere" motions and rearrangements of electrons and nuclei. However, there is one very notable exception to this, even in everyday life. It involves a thoroughly familiar phenomenon interpreted in a modern light, namely, light! A beam of light is nothing but an assemblage of massless particles, *photons*, moving at (what else?) the speed of light. Because they are massless, photons are easy to create. They are created whenever the light switch is turned on. Regarded microscopically, what happens is that they are produced in electron and atomic collision processes taking place in the light source when the latter is heated or otherwise "excited." Photons are destroyed when they impinge on and are absorbed by nontranslucent material bodies (walls, books, the retina of the eye, etc.).

Photon creationism-destructionism actually entered the world when Einstein proposed his particle-like interpretation of electromagnetic radiation. But the photon concept had a protracted birth, and the photon is anyhow such a special particle. It is massless; it is the quantum of a field we have known classically. Somehow, for photons, the enormity of creation-destruction as such did not seem to attract much philosophical discussion in the early years of this century. In any case, for a while one could still cling to the idea that "real" *ponderable* particles, particles with nonzero mass such as electrons, protons, and neutrons, are truly immutable. But there is no such immutability for them either. This first became apparent with the discovery of the neutron and the recognition of its role in nuclear beta decay. The basic beta decay reaction is

$$\text{neutron} \rightarrow \text{proton} + \text{electron} + \text{antineutrino}.$$

The neutron is destroyed, the proton, electron, and antineutrino created. The antineutrino, which is highly unreactive, easily escapes the nucleus and passes through the earth, the solar system, the galaxy, and into outer space without leaving much of a scratch. But that's another story.

Where does quantum theory fit in? The quantum theory of the electromagnetic field got its start in the heroic period of the mid 1920s when the foundations of quantum mechanics were being established. Quantum electrodynamic theory was designed from the beginning to account for photon creation and destruction. The photon emerges naturally in the theory as a quantum of the electromagnetic field. Since that time physicists have brazenly invented other fields, fields not known to us in their classical guise but that are invented for the purpose of being quantized to yield other particles as well. So, for example, there is a field that makes and destroys electrons. The older theories used to have separate fields as well for protons, neutrons, pions, and so on. We have now reached a more basic level involving, among other entities, quarks and gluons. But these too can be created and destroyed.

Beginnings

In its modern form, the structure of quantum theory was laid down in the middle of the 1920s in a concentrated burst of creativity and transformation that is perhaps without parallel in the history of scientific thought. Mainly, the creators were very young: Werner Heisenberg, Paul Dirac, Pascual Jordan, and Wolfgang Pauli were all in their twenties. The elders included Erwin Schroedinger, who published his famous wave equation at age thirty-nine, and Max Born, who at the age of forty-three recognized and helped elaborate what Heisenberg had wrought. The new outlook brought with it an unintuitive concept of reality along with a number of attendant oddities of various sorts. Among contemporary physicists, some could not readily absorb the new doctrine. They grumbled and fell out. But already the earliest applications to phenomena met with convincing success. Informed dissidents, Albert Einstein foremost among them, soon accepted the effective correctness of quantum mechanics. They were reduced to hoping that classical reality prevails at some deeper level of nature not readily accessible to observation. That deeper level, if there is one, is still today nowhere in sight. As far as the eye can see, the principles of quantum mechanics stand irreducible and empirically unchallenged. In cases where the difficult experiments and corresponding theoretical calculations can be carried out with high precision, quantitative agreement is spectacular. As often happens in intellectual revolutions, it was the younger generation that could adapt to the new ways of thinking somewhat more easily than the older one. Succeeding generations have had an even easier time of it; they simply grew up with the subject. Nevertheless, the world view of quantum mechanics is odd; and the oddest thing of all is that, still today, many decades after its foundation, quantum mechanics continues to seem odd even to scientific practitioners who work with the subject every day and who know and operate confidently in its framework. Their wonderment expresses itself not so much at the operational level as at a philosophical one. Deep questions persist at that level. We will surely not resolve them here. The more modest aim here is simply to convey some notion of what quantum mechanics is: its principles and some of its consequences and oddities.

Many questions within the classical framework were still unresolved toward the end of the nineteenth century, especially questions having to do with the nature of atoms—and for some diehards, even the very existence of atoms. But the Newtonian framework was not in doubt. It is possible today in hindsight to recognize hints of quantum effects, empirical departures from classical expectation that should have been pounced on by our nineteenth century ancestors. However, this is only in hindsight. They *did* in fact encounter anomalies and *did* fret over them, but it was far from clear at the time that these could not be resolved within the still developing classical picture.

There are vast stretches of contemporary macroscopic science and engineering that still do very well today without any reference at all to the quantum mechanical basis of nature. This is so because classical Newtonian behavior emerges for the most part as a very good approximation to quantum mechanics for macroscopic systems. But this assertion has to be understood in a qualified sense. The qualification can be illustrated by means of an example. Consider the flow of oil through a smooth cylindrical pipe, the flow being driven by a pressure differential that is established between the ends of the pipe. If the pressure differential is not too large the flow will be smooth; and it is then an easy matter, a standard textbook problem in classical fluid dynamics, to compute the flow rate, the volume of oil transported per unit time. The answer depends on the length and diameter of the cylinder and on the pressure differential. These are parameters of experimental choice or circumstance. But the answer also depends on the viscosity of the oil. If the value of that parameter is simply accepted as a given fact of nature, as a quantity to be determined empirically, then the computation of flow rate may be said to proceed along purely classical lines without reference to quantum mechanics. However, to understand why oil has the viscosity and other properties that it has, one has to move down to the atomic level. And there the differences between quantum and classical science are as striking as can be.

Another qualification should be noted. The quantum mechanical rules, the concrete equations, are definite and well established. *In principle* one can compute the structure of oil molecules and work out the way these molecules interact among themselves in bulk oil and thence go on to the viscosity of oil. But a completely detailed calculation that traverses the whole route from the individual molecule and its ingredients all the way up to the astronomical number (about 10^{24}) of molecules present in even a small drop of oil is utterly unthinkable. The single molecule is already complicated enough. Thus, approximations and aggregate treatments have to be adopted along the way, relying on various rich and active fields of scientific inquiry; for example, the field of statistical mechanics. A pumper who wants highly accurate predictions of flow rate is well advised to adopt the empirical value of viscosity. But that same pumper may also share with others a curiosity about why things are the way they are. Moreover, there is the possibility of learning enough at the microscopic level to design molecular additives that can alter the viscosity in wanted directions.

As with viscosity, so too for other kinds of information that enter in parametric form into the various branches of classical science and engineering: tensile strength of materials, thermal conductivity, electrical resistance, equations of state (the relation of pressure to density and temperature) for various gases and liquids, optical reflection coefficients, and so on. The different fields have their independent methodologies and concepts. None suffers any shortage of engaging intellectual and practical challenges within its own framework. But so far as we know, science is seamless. At a deeper level the different fields share in common the science of atoms, where the quantum reigns. Deeper still is the fantastic world of the subatomic particles; and farther out, the world of the cosmos.

Quantum mechanics first began to intrude itself on mankind's attention in the very first year of the twentieth century. It did not by any means spring up full grown. The beginnings can be sharply placed within a rather esoteric corner of the scientific scene of those times; namely, the physics of *blackbody radiation*. The blackbody question has to do with the frequency spectrum of electromagnetic radiation that fills any volume of space surrounded by material walls in thermal equilibrium. That seems an awfully specialized topic. However, it had been established decades earlier through elegant thermodynamic reasoning that the spectrum, the radiation intensity as a function of frequency, must be of a fundamental character. It can depend only on frequency and temperature, not on the shape of the vessel nor, more strikingly, on the kinds of materials that the walls are made of. Deep issues therefore appeared to be at stake. Experimental measurements over various parts of the frequency spectrum were actively pursued toward the end of the century. The challenge on the theoretical side was to predict the spectrum. It was the German physicist Max Planck who succeeded. That was in the fall of 1900. We will describe the scientific issues more fully later on; but briefly, what happened was this. Presented with the latest experimental results on the blackbody spectrum, Planck sat down at one point and in not much more than an evening's work so far as we know, he devised—stumbled upon—an empirical formula that fit the spectral data remarkably well. This was something more than a case of raw curve fitting, however, since he brought to the task some guiding ideas that had emerged from earlier work by himself and others. Nevertheless, his formula was essentially empirical. Over the succeeding months he sought to deduce it within the framework of the classical theory of his times. This required some statistical mechanics reasoning. But the statistical mechanics aspects of classical science were still somewhat in flux and Planck did not recognize, or at any rate did not choose to follow, a simple path to the blackbody spectrum that was available to him. Had he taken that path (noticed slightly earlier by Lord Rayleigh), he would have encountered catastrophic disagreement with the data. Instead, he followed a more complicated route that was mostly classical in its outlines, but then did some fiddling that we will describe later on. Out came the empirical Planck blackbody formula! From this small seed came the quantum revolution.

There was no immediate commotion in the streets. Only a small band of scientists were participating in or paying close attention to these developments. Among those few it was pretty clear that something new was afoot, but it was far from clear what that new thing was. A decisive insight was provided by Albert Einstein in 1905, the miracle year in which, among other things, he published his papers inaugurating the special theory of relativity. What Einstein drew from Planck's discovery was the startling hypothesis that electromagnetic radiation of frequency f can exist only in discrete energy bundles, *quanta*, and that the energy of each such bundle is proportional to the frequency: energy $= hf$, where the proportionality constant h is the new parameter of nature that had entered into Planck's blackbody formula. These quanta of Einstein are particle-like entities that have since come to be called *photons*. However, light is nothing but a form of electromagnetic radiation; and one of the triumphs of nineteenth century science had been the discovery that light is a wavelike phenomenon. Here then, with Einstein's quanta, was the beginning of the celebrated wave-particle duality conundrum that hovered over physics during the next two decades.

Quantum ideas were soon extended from radiation to ponderable matter. In fact, Planck's work had already suggested some sort of energy quantization for ponderable matter; but, excusably for that pioneering effort, the suggestion was rather murky. Following up on these hints, in 1907 Einstein developed a simple quantum model of the specific heat of material bodies. Specific heat is a parameter that characterizes the temperature change induced in a material body when it absorbs a given quantity of heat energy. Einstein proceeded as follows. Material bodies can of course sustain sound waves over some range of frequencies f. For these he adopted the same quantization hypothesis that he had adopted for electromagnetic radiation; namely, the assumption that the energy in a sound wave disturbance of frequency f can come only in bundles of energy hf. He was content to take a single representative frequency. Others soon generalized to cover the whole frequency range. The model provided a qualitatively successful account of certain anomalies, departures from the expectation of classical theory, that had been known empirically for some time. The band of scientists paying attention to quantum developments began to grow.

In 1913 the young Danish physicist Niels Bohr turned to the inner workings of the atom. What might the developing quantum ideas have to say on this subject? For the content and structure of the atom he took up a model that had been convincingly proposed only a couple of years earlier by the great experimentalist Ernest Rutherford. In it the atom is pictured as a kind of miniature solar system: a tiny, positively charged nucleus at the center (analog of the sun), and very much lighter, negatively charged electrons (the planets) orbiting around the nucleus. Rutherford came to this picture of the atom through a celebrated experiment in which his colleagues H. Geiger and E. Marsden bombarded a thin metal foil with fast alpha particles and observed, to their wonderment and Rutherford's, that the alpha particles occasionally scattered through large angles. Collisions with the atomic electrons, which are very tiny in mass, could not be expected to produce substantial deflections of the fast, heavier alpha particles. But a heavy, highly concentrated positive charge, an atomic nucleus, would do the trick. On this picture, Rutherford could work out the expected distribution of scattering angles, proceeding along classical Newtonian lines based on the Coulomb law of force between charged particles. The result agreed well with experiment and confirmed Rutherford in his model of the atom.

But the Rutherford atom presented a conundrum. To illustrate, consider the simplest atom, hydrogen. It has a single electron moving around a proton nucleus. The electron, acted on by the Coulomb force of the nucleus, is in a state of accelerated motion. According to the classical laws of electricity and magnetism, an accelerating charge must constantly be emitting electromagnetic radiation and thereby losing energy. Suppose for a moment that this energy loss can be ignored. Then, classically, the electron travels in an elliptical orbit with a revolution frequency that depends on the electron energy among other things. It radiates at the frequency of that orbital motion. But there are infinitely many *possible* orbits, just as in the case of objects (planets, comets, asteroids, spaceships) moving around the sun. Given a macroscopic collection of hydrogen atoms, it would be surprising if the electrons in the different atoms were not traveling in a whole range of different orbits. That is, on this picture one would expect an essentially continuous spread of radiation frequencies. In fact, however, atoms radiate only at certain discrete frequencies, in a characteristic pattern that distinguishes one species of atom from another (one speaks of the characteristic frequencies as "lines" since they show up as lines in a spectrographic display). An even more serious problem for the classical Rutherford atom is that one is not really allowed to ignore the fact that the electron is losing energy as it radiates. Instead of traveling steadily on an elliptical orbit, therefore, a classical electron must eventually spiral into the nucleus, its orbital frequency and thus the radiated frequency changing all the while as the orbit shrinks in size. Empirically, however, nothing like this makes either spectroscopic or chemical or common sense. Confirmed atomists had in fact been confronted with these paradoxes for a long time, trying to figure out how it is possible, classically, to stabilize atoms against radiative collapse; also, how to account for their discrete line spectra.

Here, presented in a series of steps, is what Bohr did to resolve the conundrum, at least for the one-electron atom. Step 1: Ignore radiation for the moment and work out the electron orbits using purely classical dynamics, as discussed above. Bohr restricted himself to circular orbits. Step 2: Now impose a "quantum condition" devised by Bohr to determine which orbits are quantum mechanically "allowed," all others simply being forbidden! A consequence of this will be that only certain energies are possible. Instead of spanning a continuous range of possible values the allowed energies now form a discrete set; they are *quantized*. Step 3: Assert that the electron does not radiate while moving in one of these allowed orbits. But when the electron happens to be in an excited level of energy E and "decides" to jump to a lower level of energy E', it emits a photon of frequency f determined by the equation $hf = E - E'$. This equation is arranged to insure energy conservation, since according to Einstein hf is the photon energy.

Bohr invented his rules very soon after learning of a remarkably simple empirical formula that the Swiss schoolteacher, Johann Jakob Balmer, had devised many years earlier for the frequencies of the hydrogen atom. Balmer's formula, which involved only a single adjustable parameter (the "Rydberg"), predicted that there should be infinitely many hydrogen lines. Only several of the lines were known in Balmer's time, many more when Bohr turned to the subject. There can be no doubt that Bohr tailored his quantum rules to fit the facts. But the remarkable thing is that he *could* fit the facts, that his simple but classically inexplicable rules worked. Bohr could determine the Rydberg solely in terms of basic parameters that were already known and over which he had no freedom to make adjustments; namely, the charge and mass of the electron, and Planck's constant h. The agreement with experiment was very good indeed.

A vigorous and greatly broadened era of quantum theory now got under way as physicists sought to expand Bohr's beachhead to cover the effects of external electric and magnetic fields on the energy levels of hydrogen, to incorporate relativistic effects, to apply quantum ideas to multielectron atoms, and so on. Bohr's quantum conditions were speculatively generalized to cover this wider range of questions. Just as in Bohr's original formulation, the generalized rules had an ad hoc character: quantum conditions superimposed on top of classical reasoning without any deeper understanding of where those quantum conditions come from. To a considerable extent, developments were guided by the so-called *correspondence principle*, which had been formulated and exploited by Bohr and then taken up by others. Roughly, it is the notion that quantum behavior must resemble classical behavior for large energies. This idea was adopted and then ingeniously (and nervily) pressed into service for all energies. There were failures, but there were also many successes. It was a zany era of progress and confusion, a hodgepodge of inexplicable quantum rules and classical dynamics. It flourished for about a dozen years, the interval between Bohr's 1913 papers and the birth of modern quantum theory. The physicist Isidor Rabi, looking back, described it as a time of "artistry and effrontery."

The modern theory began along two seemingly unrelated lines, one opened up by Heisenberg, the other independently by Schroedinger. The pace was breathtaking. The first steps were taken by Heisenberg on a holiday in the spring of 1925. Although constrained and indeed guided to some extent by the correspondence principle, he broke sharply with the concepts of classical mechanics at the atomic level. He argued for abandoning the notion of definite positions and momenta on the ground that these are basically unobservable at that microscopic level. But atomic energy levels *are* observable through their role in determining the frequencies of atomic lines. Heisenberg set up a new mechanics aimed at that target. What he postulated seemed to come out of the blue; and it was expressed in a mathematical language that was unfamiliar to many, even to Heisenberg himself. However, it had the air of being on the right track. Heisenberg's mentor at Göttingen, Max Born, received the paper favorably, puzzled a while over the mathematics, then recognized it for what it was. Within a few brief months, by September, he and another assistant, Pascual Jordan, completed a paper extending Heisenberg's ideas and identifying his mathematical objects as *matrices*. The story is told—if true, it says something about the times—how the then unknown Jordan came to work with Born. The young man found himself traveling in a railroad compartment with Born and a colleague of Born's. Born was talking to his colleague about matrices. Jordan overheard, introduced himself, and said that he knew about matrices and maybe could help. Born signed him on, just like that! Their joint paper was produced not much later. Soon after, in November, Heisenberg joined Born and Jordan to produce the celebrated "three-man" paper (*Dreimanner Arbeit*) which set out in an extended and logical framework Heisenberg's quantum theory, now dubbed *matrix mechanics*. Meanwhile, basing himself only on Heisenberg's original paper and unaware of the work of Born and Jordan, Paul Dirac in Cambridge similarly extended Heisenberg's ideas, in a different, elegant mathematical language. It brought out the formal similarities between quantum and classical mechanics, and also the differences. Before the year was out Pauli had already applied the new quantum theory to the hydrogen atom. In particular, he successfully worked out the effect of an electric field on the energy levels of hydrogen, a problem that could not be tackled in the old quantum theory.

All of this was in the space of not much more than half a year! And then, in the very first month of the next year, 1926, there came the first of Schroedinger's papers laying out what looked to be an entirely different quantum theory. Schroedinger built on an idea that had been introduced several years earlier in the doctoral dissertation of Louis de Broglie, who was by then almost an elder at age thirty! What de Broglie suggested was that just as light had been shown to be both wave-like and particle-like, so too perhaps there are "matter waves" somehow associated with ponderable matter, for example, electrons. Einstein recognized the promise in this idea and gave it his influential blessing. Schroedinger extended it into a full-blown theory. Pursuing analogies with classical mechanics and optics, he introduced the idea of a wave function that is to be associated with any system of material particles; and he wrote down an equation that the wave function must satisfy, all of this even though the physical meaning of this function was initially quite vague. No matter that it was vague, however. The equation passed a first and by now mandatory test. It produced the right energy levels for the nonrelativistic hydrogen atom. Except for some initial reserve, even grumpiness, on the part of Heisenberg and others at Göttingen, Schroedinger's papers quickly captivated the world of physics. Unlike matrix mechanics, his wave mechanics was expressed in a familiar mathematical language; and, initially, it had about it the air of a theory that might be reconciled with classical notions of reality. That latter proved to be an illusion.

If a vote had been taken at the time to choose between the two theories it is probable that most physicists would have boycotted the election altogether (a pox on both of these newfangled quantum theories!). Among the voters, however, it is likely that the majority would have opted for wave over matrix mechanics. But it soon transpired that these two theories are really one and the same, as Schroedinger could demonstrate convincingly enough and as others could soon prove to higher standards of mathematical rigor. The two theories, that is, are just two different mathematical *representations* among an infinite number of other, possible representations of the same physics. This is not altogether unlike the case of different coordinate systems being used to describe the same phenomena but from different vantage points. The principles of quantum theory can in fact be formulated in highly abstract terms that do not commit to any particular representation. However, both for practical calculations and for purposes of developing an intuitive feel for quantum mechanics, it is usually best to come down from the abstract heights. It will be most convenient in the present exposition to proceed along the Schroedinger line.

Quantum mechanics was taken up widely and quickly following the papers of the founders. The earliest applications concentrated on various energy level problems. It was possible to address this class of problems without facing up to broader interpretative questions; in particular, questions having to do with the physical significance of the Schroedinger wave function. The modern interpretation was supplied soon enough, however, beginning with a remark made by Born in a 1926 paper on the quantum theory of scattering. This was swiftly elaborated. Above all others, it was Niels Bohr who presided over development of the general interpretative principles of quantum mechanics. What emerged was the picture of a probabilistic structure of nature and hence a sharp break with intuitive

notions of reality. Among the giants, Schroedinger himself resisted, as did Einstein. Einstein watched with "admiration and suspicion." For a time he pressed his antiprobabilistic outlook ("God does not play dice") in a celebrated series of debates with Bohr. Bohr won. Einstein eventually accepted the correctness of quantum mechanics as far as it goes; but for the rest of his life he held out for the existence of a deeper, not yet accessible, level of classical reality.

What does the wave function signify? Everything. According to the principles of quantum mechanics the wave function incorporates all that can be known about the state of the system at any instant. But it does not in general tell where the particles are located or what their momenta are. What it gives us, and that's all we can know, are *probabilities* concerning the outcomes of various kinds of measurements that might be made on the system, measurements of position, momentum, energy, angular momentum, and so on.

The contrast with classical language is interesting here. For example, a classical scientist will write "let x denote the position of the particle," rather than "let x denote the outcome of a *measurement* of the position of the particle." Classically, unless one is concerned with the practicalities of a measurement, it will be understood that the particle surely *is* somewhere. Yes, its position variable *can* in principle be measured, but there is no need to emphasize the latter point or speak of measurement. Quantum mechanically, on the other hand, the particle is *not* at some definite place, not unless a measurement reveals it to be at that place. One can speak only of probabilities in connection with a measurement of position and other variables. The notion of measurement, therefore, is nearer to the surface in quantum mechanics. Heisenberg: "We can no longer speak of the behavior of the particle independently of observation." Bohr: "An independent reality can neither be ascribed to the phenomena or the agencies of observation." Three baseball umpires: First umpire, "I calls them the way I sees them." Second umpire, "I calls them the way they *are*." Third umpire, "They ain't nothing till I calls them."

Let us return briefly to the historical story. Schroedinger's version of quantum mechanics brought out clearly the wave-particle duality aspect of ponderable matter. Wave-particle duality for electromagnetic radiation, whose particle-like aspect is the photon, found its proper quantum basis in 1927 with the application of quantum principles to the electromagnetic field. This was the work of Paul Dirac, who inaugurated *quantum electrodynamics* in a paper published that year. Dirac struck again in the following year, 1928, with his relativistic wave equation of the electron. Apart from an unsuccessful early attempt to marry his quantum ideas to special relativity, Schroedinger's quantum theory had addressed itself to nonrelativistic situations, situations where velocities are small compared to the speed of light. Dirac succeeded in constructing a relativistic quantum theory of the electron, a theory that incidentally (!) predicted the existence of antiparticles—although Dirac did not initially recognize that implication.

By the end of 1928 the foundations of quantum theory were firmly settling in.

CHAPTER TWO

Classical Background

Newton's Law

Quantum mechanics may run against the grain of ordinary experience, but the Newtonian mechanics that it supplanted also took some getting used to by our ancestors (and still does, for many contemporaries). Probably the two most widely quoted mantras taken from physics are Einstein's $E = mc^2$ and Newton's

$$\mathbf{F} = m\mathbf{a}. \tag{2.1}$$

In this chapter, we will be looking at the world from a pre-quantum perspective; and to begin with, a nonrelativistic one as well. Newton's equation governs the motion of an object of mass m acted on by an external force $\mathbf{F}$. The concept of mass we can leave unanalyzed for the moment, beyond assuming that it is well enough captured by the reading on a weight scale. Acceleration $\mathbf{a}$ is the rate of change of velocity $\mathbf{v}$: $\mathbf{a} = d\mathbf{v}/dt$. The symbols $\mathbf{F}$, $\mathbf{a}$, and $\mathbf{v}$ are all written in boldface to emphasize that they are *vector* quantities; that is, they have not only magnitude but also direction (e.g., the car's velocity is 60 miles per hour, pointing in the northeast direction).

Many of the ancients, Aristotle among them, thought that rest was the natural state for material bodies; that motion required the influence of external agencies, forces as we would now say. But according to Newton, in the absence of forces it is *acceleration* that vanishes, not necessarily velocity. In this sense, the natural state, the state of motion if no forces are acting, is one of uniform velocity; namely, linear motion at steady speed. Rest is just a special case in which the speed happens to be zero. To be sure, our everyday terrestrial experience belies all of this. For example, stop pulling the cart and it slows to rest. But we have come to recognize that even in the absence of deliberate pushing and pulling the ground exerts a frictional force on the moving cart. Indeed, the familiar forces of everyday terrestrial life are mainly *contact* forces of various sorts: friction itself, the brief contact with a baseball bat that changes the direction and magnitude of the baseball's velocity, the push by the road that is provoked by the rotating tires and that overcomes friction and can even accelerate an automobile, and so on.

This is a good place to call attention to a supplementary law that goes along with Newton's Eq. (2.1). It asserts that the forces acting between any pair of objects, one upon the other, are equal and opposite. If body A exerts a force $\mathbf{F}(A \to B)$ on body B, then the force that B exerts on A is $\mathbf{F}(B \to A) = -\mathbf{F}(A \to B)$, where the minus sign indicates reversed direction. For example, as the baseball accelerates in some direction during its brief contact with the bat, the bat accelerates (recoils) in the opposite direction. We shall continue to speak of Newton's *law* (singular), with the understanding that Newton's *laws* (plural) are being taken together; namely, Eq. (2.1) and the supplement discussed just above.

Although contact forces are a familiar feature of everyday life, one of the most pervasive forces that enters into terrestrial (and heavenly) affairs, gravity, is evidently of a different sort. It is a noncontact force. It acts at a distance. Electric and magnetic forces similarly act at a distance. Indeed, contact interactions when regarded microscopically really reflect electromagnetic action-at-a-distance between the neighboring atoms in the two objects said to be in contact. "Contact," that is, is not to be taken too literally at the microscopic level. All the forces of nature acting between material bodies in fact act at a distance in this sense. Indeed, all the forces that are relevant for everyday science and technology, beyond the nuclear and subnuclear domain and beneath the cosmic, have already been mentioned: gravity and electromagnetism!

Gravity

Let us start with gravity. Gravitation is *attractive*. The force acting on either one of a pair of gravitationally interacting objects points in a direction *toward* the other object. The magnitude of the force between any two small bits of matter is proportional to the product of their masses and inversely proportional to the square of the distance between the bits. If the masses are m_1 and m_2 and the separation distance is r, the radial force acting along the line between the masses is

$$F = -Gm_1m_2/r^2, \tag{2.2}$$

where G is an empirical proportionality constant. The minus sign is put in to represent the fact that the force is attractive. This gravitational force law, which we owe to Newton, is expressed here in a basic form that refers to bits of matter whose dimensions are so small compared to the separation distance r that they can be regarded as geometric points. The force acting between any two bodies A and B of finite size can be obtained from this by regarding each body as made up of many small bits and adding up (vectorially!) the forces that act between every bit in A and every bit in B.

The gravitational force is very weak. It acts, for example, between two books sitting on a table. But that force is far too small to overcome friction, which simply adjusts itself to compensate for the gravitational attraction between the books and prevent their motion. It takes very sensitive laboratory experiments to detect the gravitational interaction between objects here on earth, objects whose masses are of "ordinary" size. The gravitational effect that is so pervasive in everyday life has nothing to do, then, with the weak gravitational forces acting among the various objects that inhabit the earth's surface and environs. Rather, what is a matter of daily experience is the gravitational force that the whole, massive earth itself exerts on any such object. A spherically symmetric body, which the earth very nearly is, acts gravitationally on objects outside of it as if all its mass were concentrated at the center. The gravitational force that the earth exerts on an object of mass m located on the earth's surface is therefore given by $F = -GmM/R^2$, where M is the mass of the earth and R is its radius. The force acting on an object located at a height H above the surface is found by replacing R in the above formula by $R + H$. But since the earth's radius R is so large ($R = 6{,}370$ kilometers), the change in gravitational force is small even between sea level and the altitude of Mount Everest.

It will be somewhat instructive here to consider briefly what it is that happens when one jumps, straight upward, say. Initially, if the jumper is at rest, the earth's downward gravitational force is countered by an upward contact force that the surface exerts on her feet and that adjusts itself to exactly cancel the gravitational force. When she starts the jump, her foot provokes an added contact force that exceeds the gravitational one, so that during this brief interval her "center of gravity" accelerates upward. That force disappears after contact is broken and she immediately accelerates downward owing to the now uncompensated gravitational attraction of the earth. Downward acceleration doesn't necessarily imply downward velocity, however. At this stage, it merely entails that the upward velocity is decreasing with time (that is, she is moving upward but slowing down). Her motion eventually reverses direction and she begins to move downward with ever-increasing speed. The downward *acceleration* has been constant throughout, on both segments of the journey. During the brief period of contact at the start of the jump the earth exerted an enhanced, nongravitational contact force on the jumper, as discussed before. According to Newton, the jumper exerted an equal but opposite force on the earth. Thus, while she is soaring upward the earth's center of gravity is "soaring" downward. She has knocked it off course; although, owing to its big mass, not by much. After contact is broken she continues to exert a now uncompensated gravitational attraction on the earth, so the latter slows in its downward motion, finally reverses direction, and comes back to meet her on her return. She is back on the ground, the earth is back on course, all is well.

We now turn to larger issues. We will start by considering what it is that Newton's law, Eq. (2.1), says, and what it does not say. It *does* surely say that if there are no forces acting on a body then the body does not accelerate and therefore moves only at steady speed *in a straight line* (bear in mind that even if the speed is steady in magnitude, a curved orbit implies acceleration). But insofar as an object is acted on by external forces, Eq. (2.1) does not in itself tell us much, not until we know the nature of the relevant force law; that is, not until we know how the net force exerted on the object depends on its location, and perhaps on its velocity, as it moves in the force field arising from other bodies acting on it. Equation (2.1) does not acquire serious predictive power until one has independent information about the force F that enters into that equation. It is the combination of Eq. (2.1) and the detailed force law that provides the governing equation of motion. In the case of gravity the basic force law is given by Eq. (2.2). For a group of objects interacting only gravitationally, the force on any one of them will depend on its distance to each of the others, in accordance with Eq. (2.2). The equations of motion for the various objects are therefore coupled. For example, if the system is composed of two bodies, the acceleration of A depends on the distance to B. But that distance is changing with time, not only because A is moving but also because B is moving. The motions have to be treated jointly. In the two-body example that happens to be mathematically easy. For three or more bodies matters are computationally more complicated. Nevertheless, the coupled equations and the specification of initial conditions serve in principle to determine the motions in full detail. By *initial conditions* we mean the positions and momenta of all the bodies at some one instant of time. Take the case of a planet moving around the sun, supposing for simplicity that its interaction with all the other planets can be ignored. To simplify further, neglect the motion of the sun, a good approximation inasmuch as the sun is so much more massive than any of the planets. With all of this, the equations of motion are easily solved and one finds that the planet must move in an elliptical orbit, a motion characterized by six parameters (the orientation of the plane of the orbit, its major and minor semiaxes, etc.). So far as the equations of motion are concerned, these are free parameters. They have to be determined empirically. Equivalently, the three Cartesian components of the position and momentum vectors at some one instant of time serve to fully specify a particular orbit.

But now the following question arises. In what class of reference frames is Newton's law supposed to hold? Consider an object which is far from all external influences so that one can reasonably argue that there are no external forces acting on it; hence, according to Newton, it cannot be accelerating. Suppose in fact that it is not accelerating, *as observer 1 sees it*. Now consider matters from the point of view of observer 2, who is watching the object's motion from a car that is moving at steady velocity relative to observer 1. The two observers will clearly see the object as moving with different velocities relative to their respective frames, but for both the acceleration will be zero. They agree: no force, no acceleration. However, if the car is *accelerating* relative to observer 1, then 2 will see the object as accelerating in his/her frame. It is therefore not enough to say "no force, no acceleration." More generally, whether or not there are forces acting on a body, one cannot invoke Newton's law without addressing the question, in what frames of reference is this law supposed to hold? Our ancestors concluded, essentially correctly from a contemporary point of view, that Newton's law is to be understood as holding only in a preferred class of reference frames, the so-called *inertial frames*. In its finer points this supposition is entangled with deep questions of general relativity and cosmology, but in an excellent, workable approximation, a frame of reference relative to which the distant stars are (on average) at rest is such an inertial frame. So too, then, are all other frames moving at uniform velocity relative to that one. Of course, an observer fixed on the surface of the earth is not in an inertial frame. The earth is spinning and hence accelerating relative to the distant stars. Moreover, it is moving around the sun, the sun is moving within our galaxy, and our galaxy is moving relative to the distant stars. This does not incapacitate us, however. We can work things out as they would be seen in an inertial frame, then use commonsense reasoning (so we think) to translate back into our noninertial frame. We are not surprised, for example, that a plumb bob does not quite hang straight down on the spinning earth; and we can easily compute its tilt.

Another fine point has to be raised here. Built into the gravitational force law of Eq. (2.2) is the implicit assumption that action-at-a-distance operates *instantaneously*. The law presumes that the force that B exerts on A (or A on B) at any given instant depends on the relative separation of the two bodies *at that instant*. Of course, if the two bodies are not moving this is not an issue. But if they are, so that the separation distance is changing with time, is it really so that the interaction is truly instantaneous? With the development of relativity theory early in the century it became clear that the interaction cannot be instantaneous. No physical influence can propagate faster than the speed of light. If some omnipotent agent were suddenly to take hold of the sun and shake it, the earth's orbital motion would continue unaffected for about eight minutes, the travel time from sun to earth at the speed of light. The elementary theory based on Eqs. (2.1) and (2.2) gives only an approximate account of gravitational affairs, although the approximation is quite excellent when applied to "ordinary" situations involving planetary motions, rocket trajectories, falling apples, and other familiar gravitational phenomena.

The gravitational force law of Eq. (2.2) has another feature of interest, one that looks to be incidental but that is in fact quite profound. Combining that equation with Eq. (2.1), we observe that the acceleration experienced by a body acted on by a gravitational force is independent of the mass of that body: the acceleration a_1 of mass 1 due to mass 2 is (in magnitude) $a_1 = Gm_2/r^2$. The mass m_1 has canceled out so that the acceleration of 1 depends on the mass of 2 but not on its own mass and relatedly for the acceleration of 2. This accounts for the effect discovered by Galileo in the famous experiment he conducted (or is said to have conducted) from the leaning tower of Pisa: all objects, light or heavy and of whatever composition, fall to earth with the same acceleration, to the extent that air friction can be neglected! The "inertial" mass parameter that appears in Eq. (2.1) might have differed from the "gravitational" mass parameter that appears in Eq. (2.2), the ratio varying from one kind of material to another. These might have been two independent properties of any given bit of matter. But to an extraordinary level of accuracy they are known to be the same. This equality of *inertial* and *gravitational* masses was seized upon by Einstein as a central clue that led him to his theory of general relativity. General relativity is in effect a theory of gravity in which gravitational influences appear as distortions in the geometry of space-time. It has profound implications for cosmology; and it has been tested successfully in explaining certain small departures from classical (Newtonian) predictions concerning the bending of light rays passing near the sun, the advance of the perihelion of Mercury's orbit, and other phenomena. General relativity replaces the elementary theory of Eqs. (2.1) and (2.2); and it banishes instantaneous action at a distance. But it reduces to the elementary theory in very good approximation for "ordinary" situations. That has to be so, bearing in mind that the Newtonian world view took firm root largely on the basis of its successful application to planetary dynamics.

Energy

This will be a good place for some words about the concept of energy. As has already been said, the state of a classical system of particles at any instant is fully specified by the positions and momenta of all the particles. Other quantities of interest, among them energy, are defined in terms of these. But why bother introducing such a defined quantity? What is its virtue? Actually, there are several kinds of energy, and the virtue of the energy concept is that the total energy of an isolated system is a conserved quantity. As time goes on, things are constantly changing, particles move about, energy is shifted from one category to another, but the total energy is something that remains unchanged. That's worth knowing. We are all familiar with some everyday usages of the word energy and possess some intuitive ideas about the concept. For example, there is the energy of motion, or *kinetic energy* as it is called. According to the familiar definition, the faster an object is moving the more its kinetic energy. So too, for given speed the larger the mass the bigger the kinetic energy. There's also the notion of "stored" energy, or *potential energy* as we will call it. An object held aloft has potential energy with respect to the ground. If released the thing will fall, picking up speed, converting potential to kinetic energy.

For an individual particle of mass m and velocity v (the latter assumed to be small compared to the speed of light), the kinetic energy K is *defined* according to

$$K = \frac{1}{2}mv^2 = \frac{p^2}{2m}; \qquad p \equiv mv.$$

The equation on the far right defines momentum. For a system of more than one particle, the net kinetic energy K is simply the sum of the individual contributions. To illustrate the notion of potential energy, look first at a system of two particles interacting gravitationally according to the force law of Eq. (2.2). The force is central and depends on the interparticle distance variable r. We sometimes emphasize this by denoting the force by the symbol $F(r)$, stressing that the force F depends on r. Let us now broaden this to a general central force law. Potential energy is defined such that the small difference in its values at distance r and distance $r + \Delta$, where Δ is a very tiny increment in distance, is

$$\text{change in potential energy} = -F(r)\Delta.$$

By definition, the actual potential energy $V(r)$ at separation distance r is obtained by summing up all these small changes as one goes from some reference distance to the distance r. In the gravitational case, it is customary to take that reference distance to be infinity. One then finds

$$V(r) = -\frac{Gm_1m_2}{r}.$$

The total energy E for a system of two particles interacting gravitationally is then the sum of kinetic energy K and potential energy V: $E = K + V$. More explicitly,

$$E = \frac{p_1^2}{2m_1} + \frac{p_2^2}{2m_2} - \frac{Gm_1m_2}{r}.$$

Now, so far we have merely *defined* the quantities called kinetic and potential energy (and thus the total energy); but there is no scientific substance in mere definitions. We come to substance by turning to Newton's equation and the gravitational force law. From the equations of motion that these lead to, one easily shows that the total energy E is constant in time. The particles move about and their momenta change with time; hence the kinetic and potential energies also vary with time, but the total energy defined above is conserved. Once determined by the initial conditions, it remains unchanged. This is not a principle we have imposed from the outside. It is a consequence of the equations of motion. It is something to cling to in situations that are otherwise complicated. The above example was for a system of two objects interacting gravitationally. The generalization to more than two bodies should be obvious: the total kinetic energy K is a sum of contributions from each body, the total potential energy V is a sum of contributions from each *pair* of bodies; for example, six pairs for a four-body system. In fact, energy conservation generalizes to all situations in which the forces come from a potential function that is time and velocity independent, depending only on the particle position coordinates. And energy conservation extends even beyond that. To the best of our knowledge, it is an exact law of nature.

The discussion of kinetic and potential energy seems to have left out other kinds of energy that one often speaks about—

for example, heat energy. When a speeding car is braked to rest, what has happened to the kinetic energy that it once had? The usual answer is that it has gone into heating up the brake pads, the tires, a patch of road, and so on. That's correct, but what *is* this heat energy? The answer is something like the following. Even when the car is at rest *as a whole*, the atoms and molecules of which it is composed are forever moving about and interacting with each other. So too for the atoms of the road. That is, any chunk of matter has *internal* energy, kinetic and potential, apart from the energy it might have as a result of moving as a whole or interacting with external objects. So too for the chemical energy that we often speak of, the food energy stored in a jelly doughnut, the energy stored in a barrel of hydrocarbon fuel, and the like. Here we have to become even more microscopic and look inside the molecules and atoms. On the inside we encounter the internal motions of the electrons and nuclei and the potential energy associated with the forces acting among these ingredients. In a chemical reaction in which compounds A and B react to produce C and D, there has been a rearrangement of the electrons and nuclei. If the internal energies of A and B sum to a higher value than those of C and D, the excess will be "released" in the form of kinetic energy of motion of the reaction products C and D. But this in turn is a contribution to the heat energy of the environment in which these reaction products now find themselves. Conversely, if the sum of internal energies of A and B is less than that of C and D, the reaction can only proceed by stealing from the kinetic energy of motion of A and B. So the system containing the initial ingredients has to be warm enough to supply this energy. Overall, energy is conserved.

Another word here about energy. At the microscopic level a convenient unit of energy is the *electron volt*, abbreviated eV. That is the amount of energy an electron (or other object having the electron's charge) would pick up in falling through an electrical potential difference of 1 volt. For macroscopic objects, that's not a lot of energy. But it *is* a lot when concentrated on a single electron. In picking up one electron volt of energy, an electron starting from rest acquires a speed of about 600 kilometers per *second*! In energetic chemical reactions, the energy transactions per participating atom or molecule are usually not even as much as an electron volt. Photons of visible light have energies of order a few electron volts.

Electromagnetism

Gravity is a persistent presence in everyday terrestrial life but in a rather monotonous, unvarying way. It makes things fall, but that's about it. From time to time we also become aware of other kinds of action-at-a-distance forces, electric and magnetic ones: for example, the force that magnets exert on one another or that the earth as magnet exerts on a compass needle; the static electric force that a hovering comb exerts on freshly combed hair (on a dry day); and so on. But electromagnetic influences are far more pervasive than is suggested by these humble curiosities. The electrons jostling back and forth in the filament of a light bulb exert electromagnetic forces on the electrons in the retina of a distant observer. Similarly, the electrons sloshing back and forth in the antenna of a radio transmitter exert forces on the electrons in the distant receiver antenna. Moreover, all the familiar contact forces that we spoke of earlier are not strictly contact in nature at all, not when regarded microscopically. They are manifestations of the electromagnetic forces acting between the atoms on or near one surface with the atoms on or near the other surface, and so on.

Just as gravitational forces involve the masses of interacting bodies, electromagnetic forces involve electric charges, though sometimes in a hidden way. The simplest situation is that of two *stationary*, charged particles separated by a distance r. The force acting between them obeys the inverse square law, just as for gravity. It is attractive if the charges have opposite signs, one positive, the other negative. It is repulsive if they have the same sign, whether both positive or both negative ("repulsive" is not an aesthetic judgement here; it signifies that the force acts in a direction such as to push the particles apart). The radial force is governed by Coulomb's law:

$$F = Q_1 Q_2 / r^2, \tag{2.3}$$

where Q_1 and Q_2 are the charges. Notice that the product $Q_1 Q_2$ is negative if the charges are of opposite sign, positive if of the same sign. Negative implies attraction; positive, repulsion. Here again we are supposing tiny bits of charged matter, charged *particles*. Just as in the gravitational case, there is a potential energy associated with the interaction between the two charges. It is given by

$$V = \frac{Q_1 Q_2}{r}. \tag{2.4}$$

For systems containing many charges, the force on any one particle is obtained by adding (vectorially!) the forces that every other charge exerts on it. The net potential energy of the system is the sum of the potential energies between all pairs.

Coulomb's law as written here applies only to the case where the charges are held fixed in place. If it were thought to hold for moving charges, the question would again arise whether the interaction is truly instantaneous; that is, whether the force at a given instant depends on the separation at that same instant. In the case of gravity the resolution had to await the development of general relativity. For electromagnetism the resolution came earlier, through a series of scientific discoveries and advances that culminated in the great synthesis achieved by James Clerk Maxwell around the middle of the nineteenth century.

In electromagnetism the notion of *fields*, electric and magnetic, rises to prominence. According to the field concept, the force acting between charged bodies does not act directly but is instead mediated from point to neighboring point in space through the medium of continuous electric and magnetic fields. At any instant, each particle is at some definite location, moving with some definite velocity (remember, we are prequantum in this chapter). But the field quantities are defined continuously over space and time. They act as middlemen between charged particles. Each particle is a contributing source of the electromagnetic fields filling space. The electromagnetic force acting on any given particle is governed by the instantaneous fields at its location arising from the other particles. We will denote by $\mathbf{E}(x, y, z, t)$ and $\mathbf{B}(x, y, z, t)$ the electric and magnetic field vectors at time t at the space point designated by the coordinates x, y, z. The field symbols are in boldface to emphasize that the fields are vectors; they have direction as well as magnitude.

The electromagnetic force that a given charged particle experiences at any instant depends only on the electric and magnetic fields *at its own location, at that instant*. Other particles enter the picture not as direct agents of force but as sources of the electromagnetic field. The force law as such is fairly simple. A particle of charge Q moving with velocity $\mathbf{u}$ (which may be varying with time) experiences the electromagnetic force

$$\mathbf{F} = Q\mathbf{E} + Q(\mathbf{u} \times \mathbf{B})/c, \tag{2.5}$$

where c is a parameter that turns out to be the speed of light, and where $\mathbf{E}$ and $\mathbf{B}$ are evaluated at the instantaneous position of the particle. The object inside parentheses on the right is the "cross product" of the vectors $\mathbf{u}$ and $\mathbf{B}$. It is itself a vector, pointing in a direction perpendicular to the plane defined by the vectors $\mathbf{u}$ and $\mathbf{B}$ and of magnitude $uB \sin\theta$, where θ is the angle between $\mathbf{u}$ and $\mathbf{B}$ (so the cross product vanishes if $\mathbf{u}$ and $\mathbf{B}$ are parallel and is maximal when they are perpendicular). An important feature of Eq. (2.5) to be noted is that the force exerted by the magnetic field depends not only on the particle's position ($\mathbf{B}$ will in general depend on position) but also on the particle's velocity. The magnetic field exerts no force on a charged particle at rest.

The force formula is simple enough when the fields are given. The more complicated part of electromagnetic theory has to do with determining the fields, given the instantaneous locations and velocities of the charged particles that are the sources of the fields. Qualitatively one may say the following. A charged particle always generates an electric field. If it is moving it also generates a magnetic field. At any point in space (here) and instant of time (now), the fields so generated depend on where the particle was (there) at an earlier time (then), such that light could have traveled from then and there to here and now. That's a complicated way of putting things. Maxwell's equations express the principles of classical electromagnetism in a much more elegant and mathematically precise way. They exemplify the power of compact notation: such a wide range of phenomena encompassed in such a few lines of equation. It will not be appropriate in this book to deal mathematically with the Maxwell equations. We will merely quote results from time to time when that serves the developing story.

Nevertheless, at least in the eyes of a physicist, the Maxwell equations are too irresistible not to display, if only for aesthetics. Here they are, for show. Savor them.

$$\nabla \times \mathbf{E} + \frac{1}{c}\frac{\partial \mathbf{B}}{\partial t} = 0; \qquad \nabla \cdot \mathbf{B} = 0;$$

$$\nabla \times \mathbf{B} - \frac{1}{c}\frac{\partial \mathbf{E}}{\partial t} = \frac{4\pi}{c}\mathbf{j}; \qquad \nabla \cdot \mathbf{E} = 4\pi\rho.$$

The quantity ρ is the electric charge density; $\mathbf{j}$ is the electric current density. Both will typically vary in space and time. Each charged particle contributes to the charge density. If the particle is moving it also contributes to the current density, current being nothing but the flow of charge. We will not define the various symbols beyond these remarks, other than to say that the notation has been updated from Maxwell's time and, to the cognoscenti, that we are in the cgs system of units. The symbol ∇ has to do with differentiation.

To gauge the relative strength of electric and gravitational forces, it is instructive to compare the two for the case of a stationary electron and proton separated by a distance r. The electron and proton have equal but opposite charges so that the electric force, like the gravitational one, is attractive. Each force obeys an inverse square law; see Eqs. (2.2) and (2.3). Their ratio, (electric force)/(gravitational force), is therefore the same for all values of r. That ratio is astonishingly large, approximately 10^{39}.

Gravity therefore plays essentially no role in atomic phenomena. It is too weak. It wins out over electromagnetism when we fall down only because both we and the earth are essentially electrically neutral, whereas we and the earth both have mass, especially the earth.

A stationary particle whose charge is Q_1 produces an electric field that points radially outward if Q_1 is positive, inward if Q_1 is negative. The field at a distance r is given by *Coulomb's law* for the electric field:

$$\mathbf{E} = \mathbf{n}Q_1/r^2,$$

where $\mathbf{n}$ is a vector of unit length directed radially outward from the particle. If a second, stationary charge Q_2 is placed at a distance r, it experiences the force given by Eq. (2.5), with $u = 0$. The result is seen to agree with what was anticipated in Eq. (2.3). If there are many charged particles contributing to the electric field $\mathbf{E}$, the field at any point in space is obtained by adding up vectorially the contribution from each. Depending on how the charges are distributed in space, the field $\mathbf{E}$ may be a very complicated function of position; but if so, this arises from compounding the simple formula given above. So far, all of this is for *electrostatics*, for charges at rest. The electric field produced by a moving charge is a more complicated business and is thoroughly incorporated in Maxwell's equations.

Magnetic fields arise (in part) from charges in motion; that is, from electric currents. For example, electrons flowing through a wire constitute such a current. The flow in this case is driven by an electric field directed along the wire, generated, say, by a battery. The wire itself is electrically neutral, the charge on the electrons being canceled by the charge on the atomic ions. Since the ions are essentially stationary there is a net current, or flow of charge arising from the motion of the electrons. Suppose for the moment that the current is steady in time so that we are dealing with *magnetostatic* phenomena. The current generates a magnetic field throughout surrounding space, the details depending not only on the magnitude of the current but also on the shape of the wire. For a long straight wire, the magnetic field at any location in space points in a direction that is governed by the so-called right-hand rule: Grasp the wire in the right hand with thumb pointing in the direction of the current. The encircling fingers will then point in the direction of the field. The magnitude of the field falls off inversely as the the perpendicular distance to the wire. To take another case, if the wire is wrapped in a tight helix of great length, a *solenoid*, the field inside the solenoid is nearly the same everywhere and points along the axis of the helix. On the outside of the solenoid the field is small, vanishingly small for an infinitely long solenoid. For more complicated geometries, Maxwell's equations yield more complicated field configurations.

But what about permanent magnets? For example, what about a simple bar magnet with its north and south poles? It produces a magnetic field yet no currents seem to be flowing. The explanation is that there *are* currents flowing but they are not driven by external batteries or other outside influences. Instead, there are *internal* currents within the atoms. For certain chemical elements the atom behaves magnetically like a tiny bar magnet. It is said to have a *magnetic moment*. The internal currents arise in part from motions of the electrons within the atom. These motions may sum up to constitute a net current. There is also another kind of contribution to the atom's magnetic moment. It turns out that electrons behave *intrinsically* like tiny bar magnets, independent of their orbital motion around the nucleus. Thought of classically, the electron might be pictured as a tiny charged sphere spinning about an axis. This constitutes a moving charge distribution; hence a current and an accompanying magnetic field. The field distribution is much like that produced by a real bar magnet. This picture of a spinning electron also suggests that the electron has intrinsic angular momentum, which it does. One thus speaks of the electron's *spin* magnetic moment and its *spin* angular momentum. The classical image of a spinning electron is of only qualitative merit. It must not be taken too literally, since the world is distinctly quantum mechanical at the microscopic level. Nevertheless, it is a fact that the electron has an intrinsic magnetic moment, whether or not one chooses to picture it classically as arising from a spinning body. For certain chemical elements, the orbital and spin magnetic moments add up to give the atom a net magnetic moment, so that the atom behaves magnetically like a small bar magnet. If the atomic bar magnets in a macroscopic body point in random directions, their magnetic effects cancel and the body is unmagnetized. If they can be lined up, as in a permanent magnet, the body as a whole will be magnetized.

We conclude this discussion of bar magnets by noting a parallel with a certain electric field configuration. The magnetic field in the near vicinity of a real, macroscopic bar magnet has a fairly complicated spatial distribution. But farther away, the magnetic field **B** is distributed in the same way as the electric field produced by a system of two particles of equal and opposite charge separated by a fixed distance. The electric field **E** at any point in space can be obtained by adding vectorially the contributions that each charge makes according to Coulomb's law. The resulting electric field distribution is essentially the same as that of the magnetic field outside a bar magnet. It's as if the bar magnet consists of equal and opposite *magnetic charges* at the ends of the bar, each contributing to the magnetic field according to a Coulomb-like law, but with electric charge replaced by magnetic charge. This is a useful mathematical observation although it does *not* correspond to the actual existence of magnetic charges in the bar magnet. There is no direct evidence anywhere in nature of magnetic charges; although, as it happens, there is much contemporary speculation about the possibility—about the possible existence of such *magnetic monopoles* in the cosmos.

Beyond electrostatics and magnetostatics, electromagnetism reveals its most distinctive features when the sources, the charge and current densities, vary with time. The resulting electric and magnetic fields then also vary in time as well as space. But as enshrined in the Maxwell equations, a time-varying electric field makes a contribution to the magnetic field. This is apart from the contribution coming from electric currents. Similarly, a time-varying magnetic field generates a contribution to the electric field. The fields are thus coupled together, a time variation in one field serving as a source term for the other. Disturbances set up by time-varying charge or current densities in any confined region of space thereby propagate out into empty space, traveling at precisely the speed of light. Light is nothing but an electromagnetic disturbance, as are radio waves, X-rays, and other portions of the electromagnetic spectrum. The experimental and theoretical work that went into this discovery is one of the great triumphs of nineteenth century science.

Special Relativity

Although it is not the main theme of this book, special relativity simply cannot be passed over. This is for two reasons: Its discovery early in the century so dramatically changed our views of space and time, and anyhow, it is so thoroughly joined together with quantum theory in the daily practice of particle physics. The placement of special relativity in a chapter entitled "Classical Background" may seem somewhat perverse, since by most reckonings it is most assuredly part of "modern physics," but we place it in this chapter anyhow. For us, "classical" simply means non-quantum-mechanical.

Let us start with two questions: How does an observer specify such things as the position of one particle relative to another, or the velocity of a particle, or its acceleration? And what is the relationship between the descriptions produced by observers in different coordinate frames? To specify the location of a point in space one has to provide three coordinate numbers: for example, in a Cartesian system the x, y, and z coordinates of the point. But of course these numbers have meaning only when the origin of coordinates and the orientation of the coordinate axes have been chosen. Those are arbitrary choices. Two observers using different origins and/or different orientations of their coordinate axes will attribute different coordinate values to a given point in space. There is nothing deep or contradictory in this. For the moment, suppose that the two observers are not moving with respect to one another, that they are *relatively* at rest. What will they agree on? They will agree on the *length* of a vector drawn from one particle to another. The distance between a given pair of material points is an objective quantity, independent of the location of the coordinate origin or the orientation of coordinate axes. So too for the *magnitude* of the velocity vector of a particle, or of an acceleration vector, or of a force vector, or any other vector. Of course, the two observers will really agree also on the direction in which any such vector is pointing, but their *specifications* of that direction may differ. Thus, the components of a velocity vector may be u_x, u_y, u_z for one observer and a different set u'_x, u'_y, u'_z, for the other; but the sum of squares will be the same for the two observers since the magnitude of the velocity is the same for both.

Things become more interesting when we consider observers in relative motion. Once we contemplate that, it may occur to us to ask, as we have already done earlier in this chapter, in what frame (or frames) of reference is Newton's law supposed to hold? For the present discussion, in speaking of "Newton's law" we shall suppose that the force on a particle depends only on the instantaneous distances to the other particles acting on it. This is the instantaneous *action-at-a-distance* hypothesis. As already said, at least for electromagnetism this is not realistic. We will return to electromagnetism soon enough. But for the moment, adopt the hypothesis.

Start with a special reference frame that is stationary with respect to the average distant star, a frame with respect to which as many stars in the universe are moving in any one direction as in another. For the moment we will suppose that Newton's law is exact in this special frame. From the law itself we then notice a remarkable fact. If it holds in any one frame, for example, the special frame noted above, it holds in all others moving at *constant velocity* relative to that frame. Together with the special frame, these constitute the family of inertial frames. The reasoning is as follows. Common sense suggests that two observers watching a moving particle from the perspective of their respective inertial frames will attribute the same acceleration to the particle, although different velocities. But Newton does not refer to velocity. That same common sense suggests that the distance between a particle and any other exerting a force on it will be the same as seen in both frames and hence that the force will be the same in the two frames. Thus, the two observers will agree on acceleration, on force, and surely on mass. Therefore, if Newton's law holds in one inertial frame, common sense says it must hold in the other. Within any given frame there is of course the usual arbitrary choice to be made of coordinate origin and coordinate axis orientation, but that is truly harmless here.

For the following discussion we will consider two inertial frames, Σ and Σ', with their coordinate axes oriented in the same direction, one frame moving relative to the other along the x axis. The velocity v of Σ' as seen in the Σ frame is directed along the positive x axis. Of course the velocity of Σ as seen in the Σ' frame is therefore $-v$; it has the same magnitude but is directed along the negative x' axis. Finally, let's choose the origins so that at time $t = 0$ the two origins coincide. Then, here is what our everyday intuition tells us about the relationships among the coordinates of a given space-time event as reported by the two observers, Hendrik and Albert, depicted in Fig. 2.1:

$$x' = x - vt, \quad y' = y, \quad z' = z, \quad t' = t. \tag{2.6}$$

We have included the "obvious" fact that the two observers report the same clock time for any event. Inverting these equations, one finds $x = x' + vt'$, the same form as above but with the sign of v reversed, as of course must be the case. Newton's law is clearly invariant under this "classical" relativity transformation connecting unprimed and primed space-time coordinates. Suppose that both observers keep watch on a particular moving particle. Let $\mathbf{u}$ be the velocity vector as observed in the Σ frame, $\mathbf{u}'$ in the Σ' frame. It follows from Eq. (2.6) that the Cartesian

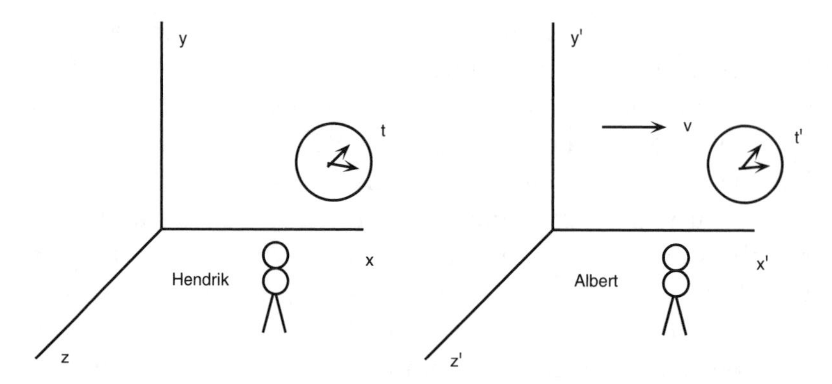

Figure 2.1 Two coordinate frames in relative motion. As seen by Hendrik (in the unprimed frame), Albert is moving to the right with speed v along the x axis. As seen by Albert (in the primed frame), Hendrik is moving to the left along the x' axis.

components of the particle's velocity as seen in the two frames are related by

$$u'_x = u_x - v, \quad u'_y = u_y, \quad u'_z = u_z. \tag{2.7}$$

All of this is simple, expected intuitively, and wrong! Not very far wrong for everyday purposes, but wrong. Questions first arose in connection with electromagnetism. The governing equations of electromagnetism, Maxwell's equations, are *not* invariant under the classical relativity transformations embodied in Eq. (2.6). This in itself need not pose a problem. Maybe position and time coordinates indeed transform as in Eq. (2.6), but maybe Maxwell's equations hold in their familiar form only in a special frame (presumably the frame at rest relative to the distant stars; or, perhaps equivalently, the frame of the ether discussed below), taking on different forms in other inertial frames. On this view it would be a piece of luck that Newton's law for velocity-independent, action-at-a-distance forces happens to have the same form in *all* inertial frames. This seemed plausible to many people, including Maxwell, in the nineteenth century. Maybe there is a subtle material medium, the ether, that fills all of space and serves to transmit electromagnetic interactions among bits of charged matter. As an analogy, consider the interactions transmitted through a body of water. Drop a rock here and now; observe the bobbing of a floating log that this produces over there, later on. The entry of the rock into the water sets up a disturbance in its immediate vicinity, the agitated water generates motions in adjacent parcels of water, and so on as the disturbance propagates outward at a speed characteristic of water waves. Perhaps there is an ether that plays the same role for electromagnetism as the body of water does for water waves, except that the ether has somehow eluded direct, physical detection. On this view, Maxwell's equations hold only in the rest frame of the ether, and only in that frame would light have the speed c predicted by those equations. In the case of water, common sense tells us that the speed of a water wave as seen by a moving observer will differ from that seen by an observer at rest with respect to the body of water. For example, if the speed of waves in the rest frame of the water is c_w, and if an observer is moving at speed v, one expects that the wave speed seen in the observer's frame will be $c_w - v$ if the observer and wave disturbance are moving in the same direction; $c_w + v$ if they are moving in exactly opposite directions; and speeds in between if the relative motions are at an angle. So too on the ether hypothesis, one would expect that the speed of light must depend on the state of motion of the observer relative to the ether.

Among the experiments bearing on these matters, the most celebrated and ultimately most decisive were the interferometer measurements first carried out by A. A. Michelson and E. W. Morley in 1887. What they discovered is that the expected effects of motion through an ether did not show up. Rather, the speed of light appeared to be universal, independent of the state of motion of the observer! Above all the others who speculated on these electromagnetic matters, it was Einstein who elevated this invariance to a basic principle, one on which he founded the special theory of relativity. His thinking doesn't seem to have actually relied very much on the Michelson-Morley findings. It had a deeper basis. Nevertheless, here is a quick, highly schematic account of that famous experiment.

Whatever the state of motion of the ether relative to the fixed stars, since the earth moves around the sun (at about 30 km/sec), it seems reasonable to suppose that it is moving relative to the ether, except perhaps at isolated moments during the year. The Michelson-Morley setup, pictured in Fig. 2.2, was designed to test for this relative motion. Light from the source strikes a half-silvered mirror *A*. A portion of the incident beam is reflected up to the mirror *B*, then reflected back down to *A* and on to the telescope. Another portion continues on to the mirror *C* and is then reflected back to *A* and on to the same telescope. The distances from *A* to *B* and *A* to *C* are the same. If the apparatus is traveling at speed v relative to the ether in the direction from *A* to *C*—call this the horizontal direction—then on the ether picture one expects the time for the round trip *A–C–A* to be

$$t_H = \frac{L}{c-v} + \frac{L}{c+v} = \frac{2L/c}{1 - v^2/c^2}.$$

For the vertical round trip *A–B–A*, taking into account that the motion of the beam in the laboratory frame is at a slant angle

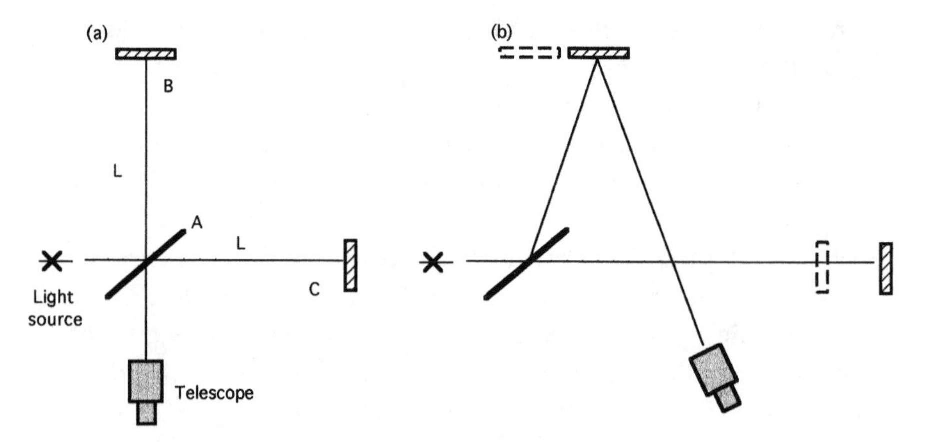

Figure 2.2 The Michelson-Morley experiment. Diagram (a) depicts the situation where the apparatus is stationary with respect to the ether. Diagram (b) corresponds to motion through the ether.

up and then down, one easily finds that

$$t_V = \frac{2L/c}{\sqrt{1 - v^2/c^2}}.$$

If the earth were at rest relative to the ether, so that $v = 0$, the horizontal and vertical time intervals t_H and t_V would be the same. The returning light waves would interfere constructively: crest would add to crest, trough to trough. But if the velocity were not zero, the times would be unequal and the interference pattern would shift. No such shifts were found. The two travel times were in fact equal, as if the speed of light is always the same independent of the state of motion of the reference frame. One way out was to suppose that the earth drags the "local" ether along with itself, hence no earth motion relative to the ether. But this runs afoul of the well-established observation of aberration of distant starlight.

H. A. Lorentz and G. F. FitzGerald found another way out. They noticed that the Michelson-Morley findings can be understood if one supposes that every bit of the experimental apparatus (and presumably any other material body) contracts by just the right amount in its dimension along the direction of motion through the ether. They were in fact on the right track and had found the right formula for the contraction, but the proposal was totally ad hoc. They could provide no physical basis for this contraction. A much deeper advance was made by Lorentz in 1904. He noticed that Maxwell's equations are invariant under a group of nonclassical transformations, of which a representative one replacing Eq. 2.6 is

$$x' = \Gamma(x - vt), \quad y' = y, \quad z' = z, \quad t' = \Gamma(t - vx/c^2),$$
$$\Gamma = \frac{1}{\sqrt{1 - v^2/c^2}}. \tag{2.8}$$

At this point, Lorentz's observation was purely mathematical. But if the equations of electromagnetism really do maintain the same form in all inertial frames, so that Lorentz transformation of Eqs. (2.8) in fact holds true in nature, then the implications for our conceptions of space and time must be quite profound.

The basic thing to focus on in any discussion of relativity is the notion of an *event*, something that happens at a particular location at a particular instant. An observer in a given reference frame associates space *and time* coordinates to an event. Observers in two different inertial frames, looking at the same event, will associate different coordinates to it; there is nothing mystifying about that. But if their clocks have once been synchronized and are in good running order, we would expect the two observers to agree on the time of the event. Indeed, we expect the connections expressed in Eqs. (2.6). What is so immediately striking about the Lorentz transformation of Eq. (2.8) is the implication that the times t' and t are not the same, that clocks run at different rates in reference frames in relative motion. Also, the connection between the space coordinates x and x' contains that unexpected, velocity-dependent factor Γ. None of this is familiar from everyday experience, but that is because the relative velocities that we normally deal with are very tiny compared to the speed of light. For $v \ll c$, the function Γ is very nearly equal to 1, and Eqs. (2.8) reduce very nearly to the intuitively comfortable Eqs. (2.6).

The Lorentz transformation recorded above is for the case where the Σ and Σ' observers have similarly oriented coordinate axes and where their origins ($x = x' = 0$) overlap at $t = t' = 0$. The Σ' frame is moving in the positive x direction at speed v relative to Σ. The transformation expressing unprimed in terms of primed quantities is obviously just the same as that given above, but with v replaced everywhere by $-v$. The algebraically interested reader can easily check this. The transformation equations for other directions of motion and other orientations of coordinate axes have the same character as for the example given here.

It was in his miracle year 1905 that Einstein founded the theory of special relativity. He based it on two very broad principles: (1) The fundamental laws of physical nature must be the same in all inertial frames. (2) The speed of light is fundamental; it must be the same in all inertial frames. This latter principle resolves the clash between Newton's law and the laws of electromagnetism in favor of the latter. The Lorentz transformation law given above emerges from these demands. Lorentz deduced his transformation law from the requirement that Maxwell's equations hold in all inertial frames. In Einstein's hands, invariance under Lorentz transformations became a guiding principle that reaches beyond electromagnetism. It has come to serve as a guide and constraint on the formulation of theories more broadly. In particular, it led Einstein to a revision of Newton's law, as we will discuss later on.

The Lorentz transformation of Eq. (2.8) relates the space-time coordinates of an event as recorded by observers in two different inertial frames. These mathematical expressions carry striking implications concerning physical measuring rods and clocks. Other entities entering into theories of nature also transform from one frame to another. For electromagnetism, invariance of Maxwell's equations requires not only the space-time connections given above but also definite relations connecting the electric and magnetic fields seen in two different inertial frames. That the fields are different in the two frames should not surprise us once we accept that the equations of electromagnetism are the same in both. For example, suppose that there is a single electric charge and that it is at rest in the Σ frame, so that there is no magnetic field in that frame, only an electrical one. In the frame Σ' moving with respect to Σ, the charge will be seen as moving. But according to Maxwell's equations, assumed to hold in the Σ' as well as the Σ frame, a moving charge generates a magnetic as well as an electric field.

The transformation formulas for the electromagnetic fields are as follows. In the Σ frame, let E_{II} be the x component of the electric field (that is, the component in the direction of motion of the Σ' frame). Define B_{II} similarly for the magnetic field. Let $\mathbf{E}\uparrow$ and $\mathbf{B}\uparrow$ be the components perpendicular to the x axis (each is thus a two-vector); and denote with primes the similar quantities in the Σ' frame. Then along with the coordinate transformations given in Eqs. (2.8), the fields transform according to the following rules:

$$E'_{\text{II}} = E_{\text{II}}, \quad B'_{\text{II}} = B_{\text{II}},$$
$$\mathbf{E}\!\uparrow' = \Gamma\left(\mathbf{E}\!\uparrow + \frac{\mathbf{v}}{c} \times \mathbf{B}\right), \quad \mathbf{B}\!\uparrow' = \Gamma\left(\mathbf{B}\!\uparrow - \frac{\mathbf{v}}{c} \times \mathbf{E}\right). \tag{2.9}$$

Having recorded this, let's now return to the space-time Lorentz transformation formulas and consider some of their strange implications.

Lorentz Contraction

Let D_r be the length of a rod *that is at rest* in the Σ' frame and that lies along the x' axis. One end is at $x' = a$, the other at $x' = a + D_r$. To find the length D_m as measured in the Σ frame, we must locate the end points *at the same instant* t in that frame. That is what it means, operationally, to measure the length of a moving object. From the transformation laws, we then easily see that

$$D_m = \sqrt{1 - v^2/c^2}\, D_r. \tag{2.10}$$

To the observer in one frame, a rod in the other frame is contracted in length (along the axis of motion). The subscript on D_r indicates that this length is as measured in a frame in which the object is at rest; the subscript on D_m indicates the length as measured in the frame moving with respect to the rod. Each observer sees a contraction of objects in the other observer's frame! This is unintuitive, to be sure, but it does not imply a contradiction.

Time Dilation

Consider two ticks of a clock that is at rest at some particular location in one frame. Those ticks take place at two different locations as seen in another frame with respect to which the clock is moving. One easily checks that the time intervals be-

tween ticks are related by

$$T_m = \frac{T_r}{\sqrt{1 - v^2/c^2}}. \tag{2.11}$$

Each observer thinks the clock in the moving system is marking time more slowly than the observer's own clock. The observer on the ground, that is, thinks his twin on the speeding spaceship is aging more slowly. The observer in the space ship similarly thinks that her twin on the ground is aging more slowly. This is the so-called twin paradox. It is not a paradox, though it *is* surprising. Both are right if the relative motion is at constant speed, the twins growing farther and farther apart. If they are to come together again to compare facial wrinkles, at least one of them will have to "turn around" and therefore undergo acceleration. The analysis of phenomena in a frame that is accelerating with respect to inertial frames carries one into the theory of general relativity. The outcome of a general relativistic analysis is that the twin who has turned back (hence accelerated) is the one who is younger when the two have come back together.

Time dilation is daily fare for practitioners of high-energy physics. Both in the cosmic rays and at particle accelerators one often deals with particles traveling at speeds very close to the speed of light. For example, consider a charged pion moving with an energy of, say, 14 billion electron volts. This is an energy that would be regarded as modest at today's frontier particle accelerators (we have chosen a figure that makes this energy 100 times the rest energy of the pion). At this energy the pion's speed differs very, very little from that of light. Now it happens that the pion is an unstable particle. It decays spontaneously to a muon and a neutrino. In its rest frame the average lifetime is 2.6×10^{-8} sec. If there were no time dilation the moving pion, which is traveling at essentially the speed of light, would on average traverse a distance of about 8 meters before it decays. Owing to time dilation, that distance becomes 800 meters! Confirmations of this sort are a commonplace nowadays.

Simultaneity

It follows from the transformation laws that events that are seen as simultaneous in one frame will not appear to be simultaneous in another frame moving with respect to it. This is yet another of the oddities of special relativity. For example, suppose, as observed in the Σ frame, that event 1 occurs at $x = 0$, $t = 0$, event 2 at $x = -D$, $t = 0$. The places are different but the times are the same so the events are simultaneous in that frame. But from the Lorentz transformation laws one easily checks that in the Σ' frame the two events occur at two different times: $t_1' = 0$, $t_2' = \Gamma D v/c^2$.

Addition of Velocities

Suppose that the observers in the two frames are watching the motion of an object whose velocity vector is $\mathbf{u}$ in the Σ frame, $\mathbf{u}'$ in the Σ' frame. From

$$dx' = \Gamma(dx - vdt) \quad \text{and} \quad dt' = \Gamma(dt - vdx/c^2),$$

one finds that

$$\frac{dx'}{dt'} = u_{x'} = \frac{u_x - v}{1 - vu_x/c^2}.$$

Similarly,

$$u_{y'} = \frac{1}{\Gamma}\frac{u_y}{1 - vu_x/c^2}; \quad u_{z'} = \frac{1}{\Gamma}\frac{u_z}{1 - vu_x/c^2}. \tag{2.12}$$

These are the laws relating the velocities recorded by two observers in relative motion. For $v \ll c$ they reduce to the commonsense relations of Eq. (2.7).

Particle Dynamics

The Newtonian force law of Eq. (2.1) can be written in the form $\mathbf{F} = d\mathbf{p}/dt$, where $\mathbf{p} = m\mathbf{u}$ is the nonrelativistic momentum of the particle and $\mathbf{u}$ is its velocity. The relativistic generalization of Newton's law was provided by Einstein. It turns out that the above Newtonian relation between force and rate of change of momentum continues to hold, but with a revised expression for momentum:

$$\mathbf{F} = \frac{d\mathbf{p}}{dt}, \quad \mathbf{p} = \frac{m\mathbf{u}}{\sqrt{1 - u^2/c^2}}. \tag{2.13}$$

With this definition of momentum, an important result familiar in nonrelativistic dynamics continues to hold relativistically; namely, the total momentum of a system of particles is conserved (is constant in time) if there is no net external force acting on the system. Individual momenta will change because the particles exert forces on one another, but the total momentum remains constant.

Einstein also generalized the definition of energy to include the notion of *rest energy*. For a free particle he took the combination of rest energy and motion energy to be

$$E = \frac{mc^2}{\sqrt{1 - u^2/c^2}} = \sqrt{(mc^2)^2 + (cp)^2}. \tag{2.14}$$

Equivalency of the two terms on the right follows from Eq. (2.13). For velocities u small compared to the speed of light, these equivalent equations reduce to

$$E \approx mc^2 + \frac{mu^2}{2} = mc^2 + \frac{p^2}{2m}, \quad \text{with } \mathbf{p} \approx m\mathbf{u}. \tag{2.15}$$

The term $p^2/2m$ is the usual nonrelativistic formula for kinetic energy. The term mc^2 is by definition the *rest energy* associated with the mass m. Thus, for a particle at rest we have Einstein's famous formula $E = mc^2$. For later discussion, and whether for low or high energies, we define the kinetic energy ("motion energy") by $K = E - mc^2$. It is the energy above and beyond rest energy. The total energy of an isolated system of particles of course includes potential energy as well as motion and rest energies. But when the particles are far apart and thus not interacting, the total energy is just the sum of the energies defined in Eq. (2.14).

We will not go through the reasoning that led Einstein to the above relativistic definitions of momentum and energy or to the relativistic generalization of Newton's law discussed above. But some further comments are in order concerning that famous equation $E = mc^2$, the rest energy of a stationary object. Consider a composite particle, say a deuterium nucleus. This is an object composed of a neutron and a proton moving around one another in a bound state. Let us go into the frame where this atomic nucleus is at rest "as a whole," so that although the proton and neutron are in motion their center of gravity is stationary. In nonrelativistic terms we usually think of the energy of such a composite as being composed of the motion energies of the ingredients plus their mutual potential energy. If we add to these the rest energies of proton and neutron, we get the total energy E of the nucleus at rest. From Einstein's formula we can then deduce a putative mass of the nucleus, $M = E/c^2$; and in fact it *is* the mass of the nucleus. The mass of the nucleus differs from and is in fact somewhat smaller than the sum of the masses of its ingredients. This is owing to the contribution of the "internal energy" of the system; that is, the motion and potential energies of the ingredients. If the system is bound, the potential energy is more negative than the kinetic energy is positive.

In general, the mass of a compound object (a nucleus, an atom, a molecule, a jelly bean) is not equal to the sum of masses of its constituents. In this sense mass is not conserved! In everyday affairs, and even at the atomic level, the differences are tiny beyond notice. For example, the mass of a hydrogen atom is less than the sum of electron and proton masses, but only by about one part in a hundred million. Similarly, the mass of a water molecule differs only very slightly from the sum of the masses of the two hydrogen atoms and one oxygen atom that make up the molecule (those constituent atoms themselves having masses slightly different than the sum of *their* component masses), and so on. For the deuteron, the difference is about one part in a thousand, small but quite detectable.

Relativistic kinematics comes fully into play in the daily business of particle physics. For example, consider the decay of the so-called Σ particle into a neutron and pion,

$$\Sigma \rightarrow n + \pi.$$

There is no profit in thinking of the Σ particle as being a compound of neutron and pion. For present purposes, let us just take things as they are, with the parent particle undergoing destruction in this process, the daughter particles being created. Let M be the mass of the Σ particle, m the mass of the neutron, and μ the mass of the pion. Suppose that the Σ particle is at rest in the laboratory frame. Denote by $\mathbf{p}$ and $\mathbf{k}$ the respective momenta of the neutron and pion when they have moved so far apart that they are no longer interacting with one another. Let the corresponding motion plus rest energies be E and ε. Suppose that we want to predict the pion energy ε. From momentum and energy conservation it follows that

$$\mathbf{0} = \mathbf{p} + \mathbf{k}; \quad Mc^2 = E + \varepsilon.$$

We have used the fact that the initial energy, that of the Σ particle at rest, is just its rest energy Mc^2. From these equations and from Eq. (2.14) one readily finds that

$$\varepsilon = \frac{M^2 + \mu^2 - m^2}{2M} c^2.$$

The masses in this example happen to be such that the meson emerges traveling pretty fast, so that the full relativistic treatment was well needed. The relativistic energy-momentum conservation laws exemplified here have been amply tested in various decay processes of this sort and in high-energy collision phenomena more generally.

Transformation Properties of Momentum and Energy

Observers in different inertial frames looking at the same particle will report different momenta and different energies. We have seen how velocity transforms from one inertial frame to another and we know how momentum and energy depend on velocity. We can therefore pretty easily work out how momentum and energy transform from one frame to another. A bit of arithmetic will reveal that $c\mathbf{p}$ and E transform in the same way that the space-time coordinates do, with $\mathbf{r}$ replaced by $c\mathbf{p}$ and t replaced by E/c, namely, one finds

$$cp'_x = \Gamma(cp_x - vE/c), \quad cp'_y = cp_y, \quad cp'_z = cp_z,$$
$$E' = \Gamma(E - vp_x). \tag{2.16}$$

The dedicated reader is invited to check that the rest energy, hence the mass, is the same in both reference frames, as of course it must be; namely,

$$E'^2 - (cp')^2 = E^2 - (cp)^2 = (mc^2)^2.$$

CHAPTER THREE

The "Old" Quantum Mechanics

Electromagnetic Waves

Electromagnetic interactions between charged particles propagate at a large but finite speed, the speed of light. Jiggle a charge over there, a charge here won't react, won't feel a force change, until the pulse reaches it. This is what gives such prominence and reality to the electric and magnetic field concepts, even though from the point of view of forces between material particles the fields may seem to be mere middlemen: charge produces field, field exerts force on another charge. Maxwell's equations are expressed in terms of these middlemen. There are infinitely many different solutions of Maxwell's equations. For example, for a pulse traveling in free space along the $+x$ direction the solutions for the $\mathbf{E}$ and $\mathbf{B}$ fields have the form

$$\mathbf{E} = \mathbf{E}_0 F(x - ct); \quad \mathbf{B} = \mathbf{B}_0 F(x - ct), \tag{3.1}$$

where c is the speed of light, $\mathbf{E}_0$ is a constant vector of arbitrary magnitude perpendicular to the x axis and $\mathbf{B}_0$ is a constant vector perpendicular to both $\mathbf{E}_0$ and the x axis. In cgs units these two vectors must have the same magnitude. In Eq. (3.1) F is an arbitrary function of the argument indicated. It should be evident, just from the fact that F depends on x and t only in the combination $x - ct$, that the pulse travels at speed c to the right along the x axis, preserving its shape. There are other solutions that describe a pulse moving to the left, in the direction of the negative x axis. They have the same structure as above but with

$F(x - ct)$ replaced by $G(x + ct)$, where G is again an arbitrary function, now of the combination $x + ct$. And there are analogous solutions for pulses traveling in all other directions. The nature of Maxwell's equation is such that, given any particular set of solutions, their sum is also a solution!

Let us return to the case of propagation along the $+x$ axis and the function $F(x - ct)$ encountered there. A special case is the sinusoidal function,

$$F(x - ct) = \sin\{k(x - ct) + \phi\}, \tag{3.2}$$

where ϕ is an arbitrary "phase" constant and k an arbitrary "wave number" constant. Recall that the sine function and its derivative repeat when the argument is augmented by any positive or negative multiple of 2π. Thus, for given time t the signal repeats when x goes from x_1 to x_2 provided that $k(x_2 - x_1) = 2\pi$ (we measure angles in *radians*; 2π radians $= 360°$). The repetition distance $x_2 - x_1 = \lambda$ defines the *wavelength*; hence we identify k with the reciprocal wavelength according to $k = 2\pi/\lambda$. Similarly, for given position x the signal repeats itself in a time interval τ such that $kc\tau = 2\pi$. This time interval τ is the *period* of the oscillatory signal. Its reciprocal is the repetition frequency f. Thus, $f = kc/2\pi$. We hereby recapture the familiar high-school formula $f\lambda = c$: the product of frequency and wavelength is equal to the speed of light. In (almost) all that follows, in order to avoid too many writings of 2π we will use the so-called *circular* frequency ω, defined by $\omega = 2\pi f$. The circular frequency is 2π times the conventional repetition frequency f. The wave number k is just 2π divided by the wavelength. The circular frequency and wave number are related by $\omega = kc$.

The most general function $F(x - ct)$ describing a signal propagating to the right along the x axis is a superposition of the sinusoidal solutions given above, summed over all wave numbers, with the phase ϕ and amplitudes $\mathbf{E}_0$ and $\mathbf{B}_0$ chosen independently for each wave number (but with $|\mathbf{E}_0| = |\mathbf{B}_0|$). A completely general solution of the free space Maxwell equations is a superposition of this kind of superposition, taken over all *directions* of propagation! The radiation coming out of the sun or a light bulb involves just such a superposition, with a range of wavelengths mostly concentrated in the visible wavelength region 0.4–0.7 microns (1 micron = 10^{-4} cm). Our eyesight, of course, evolved to respond over this interval and light bulbs are designed to accommodate to our eyes, more or less. We must also remark here that electromagnetic waves carry energy; they cause material charges to jiggle and hence acquire kinetic energy. We would not be here if the sun's rays did not carry energy to the earth. Electromagnetic waves also carry momentum, though this is less familiar in everyday life. An intense enough beam of light can not only warm you, it can knock you over.

Blackbody Radiation

It has been known since antiquity that when metals and other substances are heated to high enough temperatures, they radiate visible light; the higher the temperature, the bluer the light. The reasons became at least qualitatively clear in the mid nineteenth century in connection with the developing understanding both of thermodynamics and of electromagnetism. Light is nothing but an electromagnetic disturbance that is generated by the jiggling of charges and that propagates through space. Higher temperature implies increased jiggling; hence greater radiation intensity and also, it happens, a shift toward higher frequencies. In the 1850s Gustav Kirchhoff, a master of both of the above-mentioned disciplines, reasoned his way to a critical finding. Consider a hollow vessel whose walls are maintained at some temperature T. It was expected that the walls must be capable of both emitting and absorbing electromagnetic radiation. Although the atomic picture was not well developed at that time, one knew that electric charge is somehow present in matter and that the jiggling of electric charge must lead to *emission* of radiation. Conversely, incident radiation induces jiggling, which leads to *absorption* of energy from the radiation. Reflecting a balance between emission and absorption, the hollow vessel will be filled with electromagnetic radiation, with waves moving in every possible direction and covering a whole spectrum of frequencies. By simple but ingenious thermodynamic reasoning Kirchhoff could show that the radiation intensity must be isotropic (rays moving equally in all directions) and uniform over the vessel (the same intensity here as there). More strikingly, he could also show that the spectrum of the radiation, the radiation energy density as a function of frequency, must be completely independent of the material of which the walls are made. Let u be the radiation energy density (energy per unit volume) in a unit frequency interval at frequency ω. Since u does not depend on the nature of the walls, on all those detailed material parameters that it *might* have depended on, it must be a universal function $u = u(\omega, T)$ of frequency and temperature only. Precisely because it is universal this "blackbody" spectral function must be something of fundamental interest, something to be not only tackled experimentally but understood theoretically. It took about forty years for that understanding to emerge; or rather, to begin to emerge.

As said earlier, it was the German physicist Max Planck who did it. That was in the year 1900. But first let's consider a few things that came before. Some years earlier, the Austrian experimentalist Josef Stefan had discovered experimentally that the total energy density—the energy density u integrated over all frequencies—is proportional to the fourth power of the absolute temperature T. Later, Ludwig Boltzmann could prove this on purely thermodynamic grounds. In 1893 W. Wien proved, again by a beautiful thermodynamic argument, that $u(\omega, T)$ must have the form

$$u = \omega^3 W(\omega/T),$$

where W is some function of the ratio indicated, a function that Wien could however not predict theoretically. The reasoning that led to the above formula was impeccable. A few years later Wien reasoned himself to another result, this one not quite so impeccable; namely,

$$W(\omega/T) = A \exp(-b\omega/T),$$

where A and b are unspecified constants. In the middle of 1900, Lord Rayleigh (William Strutt) revisited the whole problem, making better use of the doctrines of statistical mechanics that had been developing all the while. He came up with the disastrous result

$$u = k_B T \omega^2 / \pi^2 c^3,$$

where k_B here is a statistical mechanics parameter named after Ludwig Boltzmann. Rayleigh's result was disastrous because the predicted energy density integrated over all frequencies, the total energy density, is infinite! Rayleigh made excuses and dropped the subject.

On October 7, 1900 in Berlin, the Plancks entertained Mr. and Mrs. H. Rubens for tea. Rubens was Planck's colleague, an experimentalist who had been carrying out measurements of the blackbody spectrum. Amidst the socializing Rubens drew Planck aside and showed him his latest results. That very evening Planck, who had been brooding over the blackbody problem for some time, sat down and worked out an empirical formula that interpolated between low frequencies, where Rayleigh's formula seemed to fit, and very high frequencies, where Wien's seemed to fit. Planck's formula agreed beautifully with the data in between as well! Both Rubens' experimental results and Planck's formula were announced in Berlin within a fortnight.

Planck was a master of thermodynamics, though a conservative fellow who was rather skeptical about the newfangled statistical mechanics. To his everlasting credit he did not rest on his empirical success. He set out to derive his formula from first principles. Luckily, he seems not to have been aware of Rayleigh's disastrous result, which was unavoidable within the classical framework of the times. Planck took a more complicated path. Since the radiant energy function is independent of the nature of the vessel walls, he was free to assume that the walls consist of simple oscillators, charged particles at the ends of springs, with all possible spring frequencies represented. By impeccable electromagnetic arguments he could relate the spectral function $u(\omega, T)$ to the thermodynamic mean energy $E(\omega, T)$ of a spring of frequency ω. Had he obtained the right classical result for this latter energy, he would have ended up with Rayleigh's formula. Instead, he fiddled around, then introduced a rather arbitrary assumption which he later acknowledged was done in desperation to achieve the result he wanted. He supposed that the oscillator can take on only those energy values ε that are integer multiples of a constant times frequency: $\varepsilon = n\hbar\omega$, where n is any non-negative integer. In effect, on this model the walls could radiate and absorb only in packets of energy $\hbar\omega$. The proportionality constant $\hbar$ is what we shall here call Planck's constant. But since Planck used repetition frequency f rather than circular frequency ω, he wrote $\varepsilon = nhf$, so *our* Planck's constant $\hbar$ is related to Planck's Planck constant by $\hbar = h/2\pi$. Of course Planck did not predict the numerical value of his constant. It entered the world as a new parameter of nature. Planck's blackbody formula in our current notation is

$$u = \frac{\hbar\omega^3}{\pi^2 c^3} \frac{1}{\exp(\hbar\omega/k_B T) - 1}. \tag{3.3}$$

Fitting this to the available data he could determine both the constant $\hbar$ and Boltzmann's constant k_B. Knowing the latter he could through well-established arguments determine the number of molecules in a mole as well as the electric charge on the electron! The results were quite good. The modern value of Planck's constant is

$$\hbar = 1.055 \times 10^{-27} \text{erg-sec} = 6.58 \times 10^{-16} \text{eV-sec}. \tag{3.4}$$

The erg is the unit of energy in the cgs system. One food calorie is worth about 40 billion ergs! As noted earlier, the symbol eV stands for electron volt, another common unit. Notice that Planck's constant has the dimensions of energy $\times$ time; equivalently, of momentum $\times$ length.

As we now know, the whole universe is filled with blackbody radiation left over from the Big Bang. It has cooled down

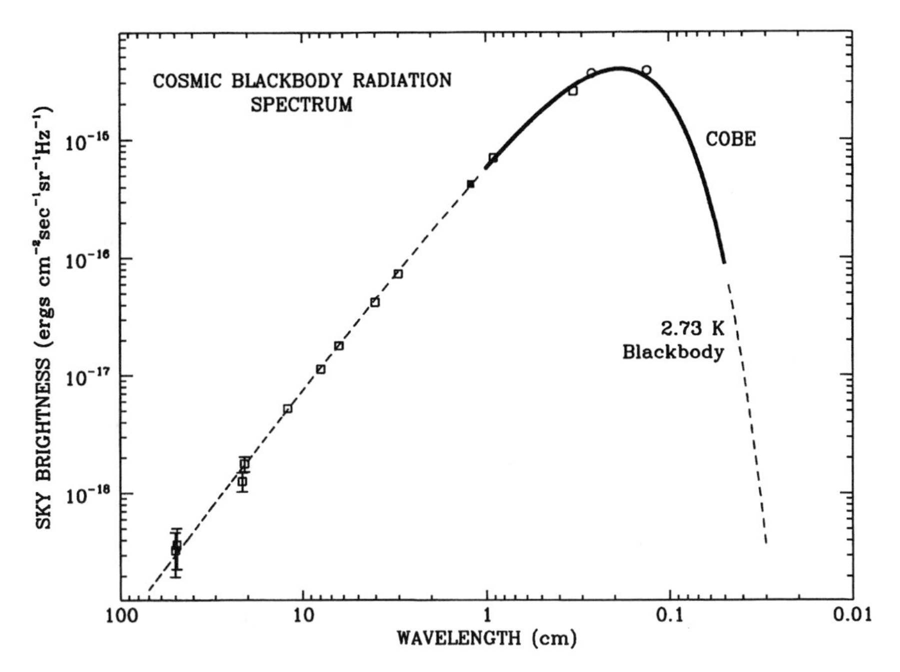

Figure 3.1 The spectrum of radiation in the cosmos left over from the Big Bang, plotted as a function of wavelength. The solid curve and boxes come from experiment (the cosmic Background Explorer, COBE); the dotted line is the theoretical blackbody curve for temperature $T = 2.73$ K above absolute zero. The fit is quite spectacular.

at our epoch to the low temperature of 2.7 degrees above absolute zero. Figure 3.1 displays the experimental findings and the theoretical blackbody curve (dotted line) corresponding to the current cosmic temperature. Closer in time to the Big Bang, the temperature was very much higher indeed.

Planck had clearly obtained a remarkably good fit to the data but it was not clear whether he had really broken new ground. Statistical mechanics was still on an unsure footing. It was Einstein above all who recognized that a revolution was at hand. Planck and others thought that the new, funny business reflected something peculiar about the interaction of charged particles and radiation. Einstein in 1905 saw more deeply. This phenomenon of energy packets, he said, is intrinsic to the radiation itself; intrinsically, radiation of frequency ω can exist only in packets of energy $\hbar\omega$. He proposed a test. It had been known for some years that charged particles come off a metallic surface when it is irradiated with ultraviolet light. J. J. Thomson had verified that these particles are electrons. The current of outgoing electrons was known to increase with the intensity of the radiation. No surprise in that. But one would have thought that the energy of the electrons would also increase with radiation intensity. Einstein said otherwise. Whatever may be the intensity of incident radiation of some given frequency, when a packet of light (photon) hits an electron, the photon can at most transfer its full energy $\hbar\omega$ to the electron. In making its way to the surface and then escaping, the electron can only lose some part of that energy. Einstein therefore predicted that the maximum electron energy, independent of the intensity of the incident radiation, is $E_{max} = \hbar\omega - \Phi$. Here Φ is the metal's work function, the energy required to escape from the surface. This "photoelectric" formula was not well tested until some years later, starting with the experiments of O. W. Richardson in 1912; then those of K. T. Compton, R. A. Millikan, and others.

Einstein's notion of energy packets was initially received with considerable skepticism, although he became increasingly respected after 1905 for his work on relativity and Brownian motion. When he was put up for membership in the Prussian Academy, his backers, including Planck, said *à propos* the packets that some excuses had to made for such an otherwise productive fellow. Einstein knew from the beginning that in a directed beam of light the packets carry not only energy but also momentum $\mathbf{p}$ of magnitude $p = \hbar\omega/c = 2\pi\hbar/\lambda$. These packets are like particles, carrying energy and momentum. As particles they seemed quite unusual: they are massless and therefore always travel at the speed of light, whatever the energy. They later came to be called *photons*. The real clincher came with A. H. Compton's 1922 paper reporting experiments on the scattering of X-rays off electrons. The scattering reaction is $\gamma + e \rightarrow \gamma + e$, where γ denotes a photon. The experimentally observed kinematics was in every respect that describing a zero-mass particle scattering off an electron.

The big trouble with all of this, however, was that light was known to be a wavelike phenomenon. How could it also display these particle-like properties? This was the great wave-particle duality conundrum. It plagued everybody who thought about it, especially Einstein.

Early Spectroscopy

It had been known since antiquity that familiar sources of light—the sun, flames, heated substances of any kind—emit a mixture of colors of light or, as we would now say, a mixture of frequencies. The rainbow is a familiar case in which the spectrum is spread out by natural means. With Newton's prisms one learned that these color mixtures can be separated at will, whatever the source of light. One speaks of the *spectrum* of radiation emitted by a source, the intensity as a function of frequency. We are not now restricting ourselves to blackbody radiation but are considering radiation sources more generally. The spectrum of any source will depend on the nature of the emitting material and on the material's condition, thermal and otherwise. In general, cold stuff doesn't radiate much at all. Radiation intensity increases with temperature. But a sample of matter can be stimulated into radiative ferment by other means as well; for example, by zapping it with an electrical spark, bombarding it with a beam of fast particles, and so on.

The spectrum will inevitably be spread out continuously over frequency, but there will also often be sharp peaks in intensity centered around certain particular frequencies. Because of the way spectral data is usually presented in charts, these peaks are called "lines." The discovery and beginning studies of spectral lines date back to the early 1800s. Actually, depending on circumstances, there can be dark lines superimposed on a continuum as well as bright ones. The bright ones represent enhanced emission at certain special frequencies. The dark ones represent enhanced absorption of radiation that is seeking to emerge from lower layers of the material. In either case, the line spectrum differs from one species of atom or molecule to another. Indeed, a new set of lines not previously known on earth was first discovered in the solar spectrum and only later identified with helium, subsequently discovered here on earth.

Early interest in spectroscopy was largely centered on its role in chemical identification and discovery. But it also seemed to some that the lines could serve as messengers from inside the atom, that they ought to have a role in telling us about those insides. The prevailing view in the nineteenth century was that the lines correspond to the frequencies of various modes of oscillation of electric charge within the atom. According to classical electromagnetism, oscillating charges can produce and absorb radiation. Relatedly, the thought was that each atom radiates at all its characteristic frequencies at once. Spectroscopists began to look at the data in a purely empirical spirit to see if they could spot any regularities in the line frequencies; for example, evidence that the line frequencies are simple harmonics of a fundamental frequency characteristic of the particular species of atom. That latter idea was not borne out.

What did prove to be a fateful discovery was made by Johann Balmer (1825–98), aged about 60 at the time, a teacher at a Swiss girls' school, a man who had never before published a single paper in physics and whose main interest seems to have been in architecture. As did others before him, he thought that the spectrum of the hydrogen atom might be the best place to look for regularities. He took up data from the work of A. Angström, who had discovered four lines in the visible part of the hydrogen spectrum and who had measured their wavelengths λ with impressive accuracy. Balmer could fit the data with the remarkably simple formula

$$\lambda = \text{constant} \times \frac{m^2}{m^2 - 2^2}, \quad m = 3, 4, 5, 6.$$

With that single adjustable constant in front, the formula worked well for all four lines. In a subsequent paper, having learned of more recent results on other lines, Balmer could get an excellent fit for lines corresponding to m ranging up to $m = 14$.

Others now took up the game for multielectron atoms, trying various formulas, with only moderate success. But in the early 1900s one idea emerged that did prove to be quite fruitful. It was suggested that one should look for formulas in which the line *frequencies* are expressed as *differences* of simple expressions. This was the idea especially of W. Ritz and came to be known as the Ritz combination principle. Actually, instead of frequency, consider what is the same variable up to a multiplicative constant, the inverse wavelength. Then notice that for hydrogen, the Balmer formula becomes

$$\frac{1}{\lambda} = \text{constant} \times \left(\frac{1}{2^2} - \frac{1}{m^2}\right), \tag{3.5}$$

a difference indeed between two very simple expressions.

The Rutherford Atom

Early in the first decade of the twentieth century, Ernest Rutherford was ensconced at Manchester, studying the passage of alpha particles through thin metal foils. Recall that the α particle is the nucleus of the helium atom. It was known in Rutherford's time that energetic α particles are ejected in the radioactive decay of certain atoms. That was interesting in its own right, but it also provided a source of energetic particles with which to bombard and thereby probe the structure of atoms. As was expected from then-current models of the atom, Rutherford found that the α particles typically scatter only through very small angles in passing through a thin metal foil. He set his colleagues Geiger and Marsden to investigate the possibility that there might occasionally be rare, large-angle events, scatterings through even more than 90°. There were such events! Not a lot, but many more than would have been expected. Rutherford was astonished. He sat down, thought, calculated, and came up with a revolutionary new picture of the atom. It was impossible, he reasoned correctly, that the large-angle events could have been caused by scattering off electrons. The electron mass is much too small to enable it to seriously deflect the much heavier α particle. The large-angle events must therefore arise from scattering off a more massive object in the atom, presumably the object that contains the positive charge of the atom. From the kinematics of such a collision he could show that in order to account for the large-angle events, the target must have a mass bigger than that of the α particle. Also, it must be very small in size, so that when the α particle comes close it can feel a strong enough repulsive Coulomb force to be seriously turned off course. Indeed, the radius cannot be much larger than about 10^{-12} cm, he deduced. All of this was for gold leaf film. The size of the whole atom was known from other considerations to be roughly 10^{-8} cm. The central, positive body—the nucleus—was therefore so small that for analysis of scattering it could be treated as a point. Rutherford worked out a formula for the expected distribution in scattering angles using purely classical dynamics. The answer depends on the charge to mass ratio of the α particle, which was known well enough, and on the charge Ze of the nucleus, which was not well known. The *shape* of the experimental curve fit the theory quite successfully. The absolute level was off. As we now know, Rutherford was off by almost a factor of 2 in the Z value for gold. But never mind. His model was a winner.

There was a remarkable piece of luck in all of this. Scattering, like all else, is governed by quantum mechanical rather than classical Newtonian laws. For most phenomena at the atomic level the two doctrines produce quite different predictions. It just so happens that for scattering in a Coulomb force field the two agree in very good approximation. Rutherford's classical reasoning produced the right scattering formula and led to the right picture of the atom. The Rutherford atom can be pictured as a kind of solar system, with all the positive charge concentrated in a nucleus that is tiny in size but that contains essentially all the mass of the atom. The electrons travel in orbits about the nucleus. The radius of the nucleus depends on the atomic species in question, but as we now know it is in fact of order 10^{-12} cm.

Bohr's Quantum Model

Despite its immediate appeal, the Rutherford atom faced some very big hurdles, as indeed did any of the atomic models that had preceded it. Let us illustrate these problems on the example of hydrogen, the simplest of neutral atoms. The nucleus of the hydrogen atom is a single proton. Its charge is balanced by a single orbiting electron.

The electron is in a state of acceleration as it moves about the nucleus, since it is continually being acted on by the Coulomb force of the nucleus. According to classical electromagnetism, an accelerating charge emits electromagnetic radiation. Suppose for a moment that we can ignore the fact that the electron must therefore be continually losing energy. We will come back to that. For the moment let it radiate, but ignore the loss of energy. One can then easily work out the orbital dynamics. The orbits are ellipses, of which the circle is a special case. The motion around an ellipse is of course periodic in time. According to classical electrodynamics, a charge undergoing periodic motion will radiate at the frequency of that orbital motion. The frequency depends on orbit parameters. But given any macroscopic collection of atoms, one would expect to find an essentially continuous range of orbit parameters. It is classically incomprehensible that the electrons would select only certain orbits and not others. It is therefore hard to see why only a discrete set of lines is observed. Anyhow, precisely because it is radiating we cannot ignore the fact that the electron is continually losing energy. This means that it must eventually spiral into the nucleus, whirling faster and faster on the way in and thereby producing a continuous spectrum. So, why don't atoms collapse? What stabilizes them? And again, why do they radiate at only certain frequencies?

Along comes the young Danish student Niels Bohr, on a stay in Cambridge to work with J. J. Thomson. Thomson had his own model of the atom, a scheme now remembered chiefly by historians. Bohr was critical of it, very politely so, but critical. In 1912 he moved on to Manchester to work with Rutherford. And there his great idea came to him. Some of the views that Bohr developed had been suggested by others around that time, but it was Bohr who had the unerring instinct for the right path.

The general view in the late nineteenth century was that the atom must have many modes of classical vibration, and that each atom radiates at all its characteristic frequencies simultaneously. But by the early years of the new century, an alternative idea was suggested; namely, that at any given moment an atom radiates at only one or another of its characteristic frequencies, and that the whole spectrum of lines from a bulk sample of atoms arises because different atoms are radiating different lines at any given moment. Bohr adopted this picture. He also firmly adopted the view that Planck's quantum must somehow enter into the atomic story. That may seem obvious in retrospect but it was not obvious at the time. After all, most of physics was going along merrily in a classical mode alongside the limited quantum forays initiated by Planck, Einstein, and a few others. But Bohr thought that the quantum must be essential for an understanding of the stability of the atom. What he did for the one-electron atom can be described in terms of the following steps.

(1) At the start, simply forbid the electron to radiate; and calculate the electron orbit on purely classical grounds. Since the nuclear Coulomb force obeys an inverse square law, the dynamical problem is the same as for the motion of planets around the sun. One knows all about the motion. The orbits are ellipses. Following Bohr, let us in particular specialize to circular orbits, where the arithmetic is especially easy. With Ze the nuclear charge and with the nucleus treated as a point particle (which it essentially is on the scale of the whole atom), the attractive radial force on the electron is $F(r) = -Ze^2/r^2$. The potential energy corresponding to this attractive force is $V(r) = -Ze^2/r$. The (inward) acceleration of a particle travel-

ing at velocity v in a circular orbit is $a = v^2/r$. From Newton's law, therefore,

(i) $$mv^2 = Ze^2/r.$$

The (nonrelativistic) energy, kinetic plus potential, is

(ii) $$E = mv^2/2 + V(r) = -Ze^2/2r.$$

The angular velocity is

(iii) $$\omega = v/r.$$

Finally, let's introduce the angular momentum $\mathbf{L}$, a vector quantity defined quite generally by $\mathbf{L} = m\mathbf{r} \times \mathbf{v}$. For a circular orbit the position and velocity vectors are perpendicular to one another, so $\mathbf{L}$ points in a direction perpendicular to the plane of motion. Its magnitude is

(iv) $$L = mrv.$$

The five variables r, v, E, ω, and L are related through the above four equations. If we know any one of these quantities, we know the others. Let us single out L and express all the others in terms of it. One can easily check that

$$r = \frac{L^2}{Zme^2}; \quad v = \frac{Ze^2}{L}; \quad \omega = \frac{Z^2me^4}{L^3}; \quad E = -\frac{Z^2me^4}{2L^2}.$$

Classically, of course, L can take on values ranging *continuously* from zero to infinity.

(2) In this step we will take some liberties with history, focusing on only one of several lines of argument that Bohr used to motivate a revolutionary "quantum condition" that he introduced. Out of the blue, more or less, Bohr postulated that L can take on only a discrete set of values,

$$L = n\hbar, \tag{3.6}$$

where n ranges over the positive integers, $n = 1, 2, 3, \ldots, \infty$. The circular orbits, labeled by the integer n, are hereby quantized by fiat! For the nth orbit it now follows that the radius, velocity, angular velocity, and energy are all similarly quantized, with

$$r_n = \frac{n^2}{Z}\left(\frac{\hbar^2}{me^2}\right); \quad v_n = \frac{Z}{n}\left(\frac{e^2}{\hbar c}\right)c;$$
$$\omega_n = \frac{Z^2}{n^3}\left(\frac{me^4}{\hbar^3}\right); \quad E_n = -\frac{Z^2}{n^2}\left(\frac{me^4}{2\hbar^2}\right). \tag{3.7}$$

The natural length in this problem is the *Bohr radius*, $a_B = \hbar^2/me^2 = 0.53$ angstroms, where 1 angstrom $= 10^{-8}$ cm. The natural energy is the *Rydberg*, $\text{Ry} = me^4/2\hbar^2 = e^2/2a_B$; numerically, $1\text{Ry} = 13.6$ electron volts. Finally, $\alpha = e^2/\hbar c = 1/137$ is the so-called *fine structure constant*. The integer n is often called the *principal quantum number*.

(3) Having ignored radiation and imposed his quantum condition to determine the allowed circular orbits, Bohr now asserted that radiation is emitted when, and only when, the electron "decides" to jump down from an orbit of energy E_n to one of lower energy $E_{n'}$. When this occurs, radiation of frequency ω_γ is emitted, the photon carrying away the energy difference:

$$\hbar\omega_\gamma = E_n - E_{n'}. \tag{3.8}$$

Conspicuously, Bohr does not tell us how and when the electron decides to jump in the process of emission. In addition to emission of radiation, there is also the phenomenon of absorption. The atom can absorb an incident photon of the right frequency by jumping *up* from one level to another of higher energy. The incident photon energy has to be just such as to supply the energy difference between the two electron levels.

Bohr's allowed states of motion (allowed orbits) are often called "stationary states," to emphasize that (by Bohr's decree) they are stabilized until the electron jumps to another stationary state. The ground state ($n = 1$) cannot radiate at all, so it is altogether stable against *spontaneous* decay. Of course, an electron in that state can jump upward if it is hit by a photon of appropriate energy. The excited states ($n > 1$) are all unstable against spontaneous decay. According to the principles of statistical mechanics, the atoms in a bulk sample of material at low temperature will mainly be in the ground state. Such a system will therefore display appropriate absorption lines but the emission lines will be weak. At sufficiently high temperatures, there will be an abundance of atoms in the various excited states, and these produce emission lines as electrons decide to jump down to lower-lying levels.

Notice that the frequency ω_γ of the photon emitted in a jump from n to n' is not equal to the frequency of either the parent or daughter orbital motion. But consider the case in which the jump is by one unit, from n to $n' = n-1$. The photon frequency is then

$$\omega_\gamma = \frac{Z^2me^4}{2\hbar^3}\left\{\frac{1}{(n-1)^2} - \frac{1}{n^2}\right\} = \frac{Z^2me^4}{2\hbar^3}\frac{2n-1}{n^2(n-1)^2}. \tag{3.9}$$

For large n, the numerator in the second factor is approximately equal to $2n$ and the denominator is approximately equal to n^4. From the third of Eqs. (3.7) it then follows that the photon frequency is approximately equal to the orbital frequency, whether that of the parent or of the daughter orbit (it does not matter which, since the two orbital frequencies are, relatively, nearly equal for large values of the principal quantum number n). This is an example of what Bohr called the *correspondence principle,* a principle that he and others exploited to guide them in the quantum thickets. Very roughly speaking, it is the notion that in the limit where allowed orbits and their corresponding energies are closely spaced on a macroscopic scale, quantum behavior ought to begin to resemble continuous classical behavior.

Bohr's theory of the one-electron atom fit the data wonderfully well, though not perfectly. One correction is easily supplied. We have treated the electron as if it moves around a fixed nucleus. In fact, both the nucleus and the electron move around their common center of gravity. This is fully taken into account simply by replacing the electron mass m in all of the above formulas by the "reduced mass" $m/(1 + m/M)$, where M is the nuclear mass and m the electron mass. The correction is very small (for hydrogen the ratio m/M is roughly only one part in two thousand), but spectroscopic data are quite accurate enough to be sensitive to this small correction.

Interest in the quantum heated up noticeably after Bohr's breakthrough, as his contemporaries sought to widen the beachhead. How was Bohr's quantum condition to be generalized in order to deal with the noncircular orbits of the one-electron atom, the effects of external electric and magnetic fields, relativistic corrections, the vastly more complicated dynamics of many-electron atoms, and so on? Generalizations of Bohr's quantum condition suggested themselves early on to a number of people and opened the way also to considerable progress with the one-electron atom. For example, Arnold Sommerfeld was able to treat the case of elliptic orbits in the one-electron atom. He generalized to two quantum conditions, with corresponding integer quantum numbers n_1 and n_2. He could then show that the semiminor and semimajor axes, b and a, are restricted in their relative sizes by the relation $b/a = n_1/(n_1 + n_2)$. The energy levels were however again given by Bohr's formula for the circular orbits, with $n = n_1 + n_2$. This implies a *degeneracy* in the energy levels, meaning that for a given value of n (hence of energy) there are as many different elliptical orbits as there are ways to partition the integer n into the integers n_1 and n_2. We will meet this degeneracy business again when we return to the hydrogen atom in the "modern" quantum context.

Progress with multielectron atoms was more spotty. However, the notion of discrete energy levels for atoms and molecules of whatever complexity became firmly established. It received striking confirmation from experiments involving the bombardment of atoms with beams of electrons. At low energies the electrons scatter only elastically; that is, the initial and final electron energies are the same. But at energies exceeding certain thresholds characteristic of the target atom, the electrons sometimes come off with reduced energy, the energy loss being offset by the energy gained as the atom changes its internal state. This could be interpreted as corresponding to collisions in which the incident electron transfers energy to the atomic system, exciting it to a higher quantum level. The interpretation was confirmed by the observation that a photon of the right frequency was emitted as the atomic system then jumped back down to its initial level.

De Broglie's Matter Waves

A critical next step on the way to the "new" quantum theory was taken by (Prince) Louis de Broglie in the midst of his thesis work in 1923. Just as electromagnetic waves had been discovered to have particle-like aspects, he argued, perhaps ponderable matter—the electron, for example—has wavelike aspects. By a piece of luck, the following line of reasoning gave some support to this conjecture. According to Einstein, the photons constituting radiation of wavelength λ have momentum $p = 2\pi\hbar/\lambda$. Now consider an electron moving in one of Bohr's circular orbits. The magnitude p of its momentum is classically a constant of the motion for a circular orbit. If there is some sort of wave associated with the electron, de Broglie said, it seems reasonable to suppose that the same relation between momentum and wavelength holds for the electron as for a photon. If so, it seems equally reasonable to require that the circular orbit accommodate that wavelength; namely, that the circumference be an integral multiple n of the wavelength. This leads to the relation $2\pi r = n\lambda = 2\pi\hbar/p$; hence $pr = n\hbar$. But for circular motion pr is the orbital angular momentum L. From this chain of suppositions he thereby deduced the Bohr quantum condition $L = n\hbar$. Einstein was impressed. He recommended approval of de Broglie's doctoral thesis.

CHAPTER FOUR

Foundations

The birth of the modern quantum theory was surveyed in Chapter 1. The pace, not just of Chapter 1 but of the events described there, was quite breathless. The foundations of quantum mechanics were pretty well laid down by 1928. Indeed, in 1926, not long after Schroedinger's first paper was published, Max Born provided the beginnings of a physical interpretation of what was going on. He presented his ideas rather casually, in a publication devoted mainly to other matters; but what he proposed represented nothing less than a revolution in our view of the world.

To get started, let us recall and then enlarge on some of the remarks made in Chapter 1 about classical dynamics. Classically one deals with two kinds of dynamical entities, particles and fields. A particle is at every instant at some particular place. A field is everywhere in space. What is "dynamical" about both is that they evolve in time. Things happen. First, consider a system of nonrelativistic point particles subject to prescribed interparticle and external forces. Let N be the number of particles in the system. The dynamical *state* of the system at any instant—namely, all that can be known about it at that instant—is fully specified by the position and momentum vectors of all the particles. Other quantities of interest, for example, the angular momentum of individual particles or of the system as a whole, the system energy, and so on, are defined in terms of the position and momentum variables. The instantaneous state is thus specified by the three Cartesian components of each position vector and three components of each momentum vector, altogether $6N$ variables, or *degrees of freedom* as they are called. The time evolution of the system is governed by Newton. The structure of his law is such that, if the state is known at any one instant it is uniquely determined for all other times.

Classically, a system of fields is a set of one or more time-varying functions defined continuously over space. The electric and magnetic field vectors are such a set. The dynamical object here is to determine the fields as a function of time for *every* location $\mathbf{r}$. This is the analog of finding the position vectors as a function of time for every particle in a system of particles. In the field case, since there is a continuous infinity of locations in space there is a corresponding continuous infinity of degrees of freedom. The field dynamics are governed by partial differential equations, for example, Maxwell's equations in the case of electromagnetism. For the electromagnetic and other kinds of field systems that one often encounters, the state of the system is fully specified at any instant by the fields and their first time derivatives, both of these being functions of the position variable. Other quantities of interest, for example, the total energy content of the field, are determined by the fields and their derivatives. If the state is known at any one instant, the governing differential equations determine it uniquely for all later times.

So much for the classical situation. Quantum mechanics introduces vast conceptual changes. In this and the next few chapters we will be concentrating on quantum ideas in the context of systems of nonrelativistic particles. There, just as classically, we continue to talk about familiar quantities such as position, momentum, angular momentum, energy, and so on. Whether in the classical or quantum context, these are examples of *observables*. An observable, recall, is a physical quantity that is in principle measurable. Although classical and quantum mechanicians talk about the same kinds of observables, the two outlooks differ enormously in what it is we can know and what we cannot know. We begin by repeating and expanding on two assertions that were laid down peremptorily in the introductory chapter.

(1) Quantum mechanically, the state of a system of N point particles is fully specified at any instant by a wave function Ψ that depends on time t and on N position vectors $\mathbf{r}_1, \ldots, \mathbf{r}_N$. *The wave function tells all that we can know about the system.* Notice, it is not the case that each particle has its own private wave function. Rather, there is a single wave function for the whole system. It depends on time and on as many vector position variables as there are particles.

(2) The wave function evolves with time deterministically, as governed by an equation that we will write down shortly. The equation is such that, if the wave function is known as a function of the position variables at any one instant, it is uniquely determined for all other times. Henceforth, we will use the terms "state" and "wave function" interchangeably.

But what does the system wave function Ψ signify? What is it that is waving? What does Ψ tell us about the expected outcome of physical measurements? The answer to these questions is a long story. It is what big, fat textbooks on quantum mechanics are all about. It is what we will be addressing more modestly as we go along.

The Two-Slit Experiment

First, let us backtrack to the origins of the wave function concept. It was de Broglie who proposed that the notion of wave-particle duality encountered for electromagnetic radiation be extended to ponderable matter. The wave nature of matter was experimentally demonstrated several years later, not long after the birth of the new quantum mechanics. The critical experiments were carried out by C. J. Davisson and L. H. Germer in the United States and G. P. Thomson in England. Here we will discuss a basically equivalent but rather idealized "gedanken" experiment that brings out the essential points. It is the pedagogically celebrated two-slit experiment. The setup is illustrated in Fig. 4.1. A source of material particles, say electrons, is located at A. The electrons are detected by an array of Geiger counters distributed along the surface C. In between, at B, there is a screen with two slits which for simplicity we will take to be identical.

First consider the case where slit 2 is closed, slit 1 open. If the electron flux emerging from the source at A is small, you will detect individual responses in the counters (individual "clicks"), just as expected on the particle picture. After many events have been recorded, you can plot the distribution of counts as a function of position x on the detector surface C. Even in a classical framework, it will come as no surprise that the distribution is somewhat spread out beyond a simple geometric projection of the slit onto C. Electrons traveling near the edges of the slit perhaps feel electrostatic forces arising from the screen; perhaps these forces produce bends in what would otherwise be expected to be straight-line trajectories. Let $P_1(x)$ be

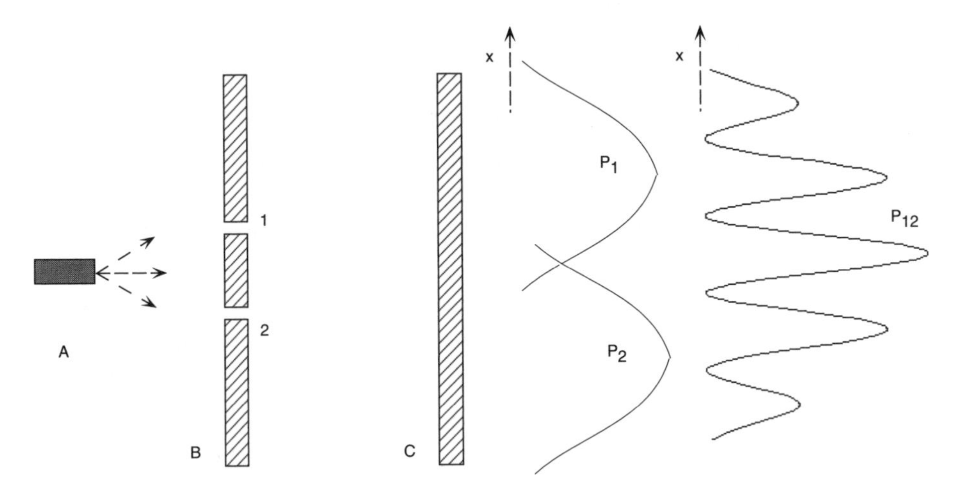

Figure 4.1 The gedanken two-slit experiment. Curves P_1 and P_2 are the count rate distributions on screen C for the respective cases where only slit 1 or only slit 2 is open. Curve P_{12} is the distribution when both slits are open.

the count rate as a function of x, the subscript indicating that it is only slit 1 that is open. Now close slit 1, open slit 2, and repeat the experiment, obtaining a probability distribution $P_2(x)$. The two distributions are shown notionally in Fig. 4.1. Classically, they depend on details that one might *in principle* take into account, for example, the velocity and angular spread of the electrons as they emerge from the source at A, those forces near the slit edges, and so on.

So far, so good. Now repeat the experiment with both slits open. Classically, the distribution $P_{12}(x)$ ought to be the sum $P_{12}(x) = P_1(x) + P_2(x)$. After all, one would think, any given electron must surely pass through one slit or the other. In fact, however, the distribution $P_{12}(x)$, shown notionally in Fig. 4.1, is not the expected sum. Moreover, its wiggly shape has the look one is familiar with in connection with wave phenomena. It is what one would expect if the source at A were a light bulb emitting classical electromagnetic radiation. In that case, we don't ask whether the light passes through slit 1 or slit 2. Light passes through both. There are electromagnetic waves everywhere; and the wave trains emerging from slit 1 and slit 2 can interfere to produce the electromagnetic equivalent of $P_{12}(x)$. A light detector such as a photographic plate on plane C responds to the *square* of the electric field E. If E_1 and E_2 are the fields associated with waves coming, respectively, from slits 1 and 2, then P_1 is proportional to $E_1{}^2$, P_2 to $E_2{}^2$, and P_{12} to $(E_1 + E_2)^2$. Notice then that P_{12} in the electromagnetic case is not the sum of P_1 and P_2; rather $P_{12} = P_1 + P_2$ plus a wiggly *interference term* proportional to the product of E_1 and E_2.

That is all very well for light, which every classicist knows is a wave phenomenon. But surely an electron is a particle; and unlike a spread-out wave, a particle that reaches the detector plane at C must have gone through one slit or the other. To check on this, let us try to catch each electron in the act of passing through one slit or the other. We shine a light on the slits and from the signal reflected by the electron determine which slit it is passing through. With both slits open, the experiment may succeed in the sense that the reflected light tells unambiguously which slit each electron went through. If so, one will find that the electrons that went through 1 will have the previous distribution P_1, those that went through slit 2 will have the previous distribution P_2; and the overall distribution independent of slit will necessarily—by definition—be the sum $P_1 + P_2$. No interference term! The act of looking has somehow changed the outcome of the experiment. But you may counter that looking involves an interaction of the electron with light waves, something that must disturb the orbit somewhat. So let us reduce the intensity of the light to minimize this. But then, sometimes the electron is not "seen" at all. For this subset of events—electrons not seen—the distribution P_{12} reverts to the wiggly curve obtained when you weren't even trying to look. In short, if you look to see where the electron is, and if you succeed, the electron is indeed at one slit or the other as it passes through the screen. But if you don't look (or don't succeed in seeing), it behaves as if it oozed through both slits, wavelike.

This gedanken experiment, which reveals the essence of numerous real experiments carried out over the years, demonstrates that electrons and other ponderable particles share with classical electromagnetism the latter's wavelike character. For material particles, the wavelike entity is the wave function Ψ. But on the other side, electromagnetic radiation shares with classical particles their particle-like character, in the sense of Einstein's energy packets for radiation. Relatedly, at low intensity of illumination from a light source at A, light detectors will record individual whole "clicks," not partial responses; and this is just what is expected of particles: Wave-particle duality!

Schroedinger's Wave Equation

As said earlier, we will be following Schroedinger's version of quantum mechanics, recognizing that it is only one of many physically equivalent representations of the abstract underlying principles. Moreover, for the present let's focus on the case of a single, nonrelativistic particle moving in some force field. Schroedinger took up de Broglie's idea that there may be some sort of wave field associated with the particle. At the start, it was still possible to suppose (just as classically) that the particle has a definite position and momentum at any moment. The new idea, however, would be that its motion is somehow guided by a wave field spread out over space (a boat drifting with the waves of the sea provides a possible image—the boat is at a particular place at any moment, but the wave disturbance guiding its drift is spread out). Schroedinger pursued various suggestive connections between classical particle dynamics and, of all things, geometric optics. These led him to a conjectured equation for a function, call it $u(x, y, z)$, that is somehow associated with a single particle of mass m and definite energy E moving in a potential $V(x, y, z)$; namely,

$$-\frac{\hbar^2}{2m}\left\{\frac{\partial^2 u}{\partial x^2}+\frac{\partial^2 u}{\partial y^2}+\frac{\partial^2 u}{\partial z^2}\right\}+Vu=Eu. \tag{4.1}$$

As a kind of mnemonic, one may relate this to the classical energy equation $K+V=E$, where K is the kinetic, V the potential, and E the total energy. In Eq. (4.1) the terms involving second derivatives may be thought of as somehow corresponding to kinetic energy. The function u is not yet *the* wave function of the particle; not necessarily, that is. We will see the connections later on. For the present, let's just see what Schroedinger did with Eq. (4.1).

Mathematically, for any given potential function V this equation has solutions, whatever the value of the parameter E. But even though the function u had not yet been supplied with a physical interpretation, right from the start Schroedinger presumed that nature accepts only those solutions u that are well behaved. "Well behaved" was taken to mean that u is bounded for all values of x, y, z and bounded as any of these variables goes to infinity; also that it is single valued, that is, uniquely determined at each point in space. With this kind of good behavior demanded, it turns out for the case of the hydrogen atom potential $V=-Ze^2/r$ that in the domain $E<0$, only certain energies are allowed; and these are just the energies that were obtained in Bohr's old quantum theory and that agree so well with experiment! For $E > 0$ all energies are allowed; the energy spectrum there is continuous, as one says.

We shall refer to Eq. (4.1), with its accompanying demand for good behavior understood, as the *energy eigenvalue equation*. The well-behaved solutions u are called *energy eigenfunctions*. The corresponding energies are the *energy eigenvalues*. Several remarks are immediately in order. The equation refers to a particle of definite energy E. Classically, no apology need be made about that. Of course the particle has a definite energy! Classically, that energy is shared between kinetic and potential in different proportions as the particle moves about, but the sum of the two is constant in time. Quantum mechanically it is going to turn out—we are not there yet—that a particle need *not* have definite energy, though Eq. (4.1) refers to the special case where it *does* happen to have a definite energy. The other thing to notice about Eq. (4.1) is that time does not enter into it. But things *do* vary with time, quantum mechanically as well as classically. The eigenfunction u has an important ancillary role in quantum theory but it is not in general the actual wave function of our particle. That wave function, $\Psi(x, y, x, t)$, depends on time as well as space.

Here is the equation that Schroedinger came up with for the actual wave function Ψ of a particle moving in a potential V:

$$-\frac{\hbar^2}{2m}\left\{\frac{\partial^2\Psi}{\partial x^2}+\frac{\partial^2\Psi}{\partial y^2}+\frac{\partial^2\Psi}{\partial z^2}\right]+V\Psi=i\hbar\frac{\partial\Psi}{\partial t}. \tag{4.2}$$

It is this that we shall call the *Schroedinger equation*. It can in no sense be said that Schroedinger derived it, or derived Eq. (4.1) for that matter, from anything that had gone before. The very thought of a wave equation was, of course, motivated by de Broglie's notion of associating something wavelike with particles. Moreover, Schroedinger could be guided by the requirement that anything he was going to come up with would have to reflect the structure of classical mechanics, at least to some extent or other. Classical mechanics, after all, works well in the everyday world. In searching for the right quantum mechanical equations, he could therefore hope to gain insight by following up on mathematical hints suggested by the classical theory. But all of this having been said, the leap of scientific imagination was stunning; the more so because the Schroedinger equation was written down before the subject of that equation, the wave function, had acquired anything other than a vague interpretation. But the truly big leap was not so much the replacement of Newton's law by the Schroedinger equation (or the Heisenberg equivalent). It was the leap to a new conception of physical reality embodied in the interpretative picture that soon followed.

Now return to Eq. (4.2). There are a number of things to be noticed about it:

(1) The imaginary number i, the square root of -1, appears in the equation. This means that we must be prepared to deal with wave functions that are complex. Recall that any complex quantity g, whether a function or a fixed number, can be broken up into the sum of a real and an imaginary part: $g = g_r + ig_i$, where g_r and g_i are real, ig_i therefore being pure imaginary. Recall that the complex conjugate of g, denoted by g^*, is $g^* = g_r - ig_i$. The *absolute square* of g is $g^*g = g_r^2 + g_i^2$.

(2) Equation (4.2) is *linear*, meaning this: If Ψ is a solution, so is $A\Psi$, where A is an arbitrary complex constant. More generally, if Ψ_1 and Ψ_2 are solutions, then so is the combination $\Psi = A_1\Psi_1 + A_2\Psi_2$, where A_1 and A_2 are arbitrary complex constants.

(3) Since the equation for Ψ involves only a first-order derivative with respect to time, it follows that if Ψ is known as a function of the spatial variables x, y, z at any one instant, it is uniquely determined for all later times. In this sense quantum mechanics is fully deterministic.

(4) No energy parameter appears in the Schroedinger equation. We may, however, notice the following. Let the time-independent function $u(x, y, z)$ be some solution of the energy

eigenvalue problem of Eq. (4.1), with E the corresponding energy. A quick check then shows that

$$\Psi(x, y, z, t) = e^{-iEt/\hbar} u(x, y, z) \tag{4.3}$$

is a particular solution of Eq. (4.2) as well as, simultaneously, a solution of Eq. (4.1). Thus, if the particle is in a state of definite energy E, its wave function Ψ is equal to the energy eigenfunction u multiplied by the time-varying exponential factor of Eq. (4.3). We may also notice more generally that if u_1 and u_2 are two solutions of the eigenvalue problem with corresponding energies E_1 and E_2, then in accordance with item 2 above the sum

$$\Psi(x, y, z, t) = A_1 e^{-iE_1 t/\hbar} u_1(x, y, z) + A_2 e^{-iE_2 t/\hbar} u_2(x, y, z)$$

is also a solution of Eq. (4.2) for arbitrary constants A_1 and A_2. But this solution involves two different energies. Which, if either, is the energy of the particle? The answer will be that the particle does not necessarily have a definite energy, or definite position, or definite momentum, or definite angular momentum, and so on! For a particle with this particular wave function, a measurement of energy can yield the outcomes E_1 and E_2, with relative probabilities $A_1^* A_1 / A_2^* A_2$.

Notice that the above solution is just a combination, with arbitrary coefficients, of solutions of the type appearing in Eq. (4.3). This obviously generalizes. The superposition of any number of solutions of the latter type is itself a solution of the Schroedinger equation.

(5) For *any* solution Ψ, the following can readily be shown. Although the absolute square $\Psi^* \Psi$ will of course, in general, depend on time as well as space, the *integral* of this quantity over all space is independent of time:

$$\int\int\int dx\, dy\, dz\, \Psi^* \Psi = \text{constant in time.}$$

A note here: when no limits of integration are explicitly shown, it is to be understood that the integration is taken over all space.

In deriving the above result one presumes that the integral is finite. This is in fact a demand of quantum mechanics, namely, that the above integral be finite, *square integrable,* as one says. If it is square integrable at any one instant of time, the above equation insures that it is so at all other times. At this point, to save some writing later on, it will be useful to introduce the concept and notation of scalar product. Given any two (possibly complex) functions f and g, define their *scalar product* by

$$\langle f|g\rangle \equiv \int\int\int dx\,dy\,dz\,f^*g. \tag{4.4}$$

Notice that $\langle g|f\rangle = \langle f|g\rangle^*$. By definition, the squared *norm* of f is $\langle f|f\rangle$; it is real and non-negative.

Probabilistic Interpretation

The collection of properties listed above suggests a first rule of interpretation. From item 3 on the list we are led to the assumption that the wave function Ψ is all that we can know about the state of the particle, in the sense that if we know it at one instant we know it for all other times. Item 5 suggests a probability interpretation. We know from item 2 that if Ψ is a solution, so is $A\Psi$ for arbitrary constant A. Let us adopt the hypothesis that wave functions differing only by a multiplicative constant really describe the same physical state; that is, the same physical situation. That being so, we might as well exploit the freedom of choice of multiplier to arrange that the wave function is *normalized,* meaning that

$$\langle\Psi|\Psi\rangle = 1. \tag{4.5}$$

Let this normalization be understood in everything that follows. Now, nothing in what has been said so far tells us anything about where the particle is. Here comes the first big step. Let us abandon the notion that the particle *is* at any particular place at each moment, replacing this with the notion that quantum mechanics deals only with probabilities. Denote by $P(x, y, z, t)$

the *spatial probability distribution*, defined such that the integral of P over any finite volume of space is the probability that the particle will be found in that volume. Following Max Born, we are led to the hypothesis that if the system is in the state Ψ, the probability distribution (probability density) is

$$P(x, y, z, t) = \Psi^* \Psi. \tag{4.6}$$

This depends not just on space but on time, since the wave function depends on both. But as is appropriate for a probability distribution, when the distribution is integrated over *all* space—over all places where the particle might be found upon a measurement—the result is independent of time and is equal to unity. We see this from Eq. (4.5) taken together with Eqs. (4.4) and (4.6).

The idea that the physical world is probabilistic in its very foundation is at the heart of the dramatic transformation produced by the quantum theory. It may be unnerving, but there it is. All that we can know about the dynamical state of a system is contained in its wave function; and the wave function does not generally imply unique outcomes concerning measurements that might be made on the system. It must be emphasized that neither Born nor any of the other founders *deduced* this interpretative picture from anything that went before. But the structure of the Schroedinger equation was at least mathematically suggestive of, and consistent with, this interpretation.

A Brief Survey of the Rules

We have been led to a probabilistic interpretation for position measurements, but that is only a beginning. What about other observables: energy, momentum, angular momentum, and so on? For any physical observable there are in fact three different questions to be asked:

(1) What is the universe of possible outcomes of a measurement?

(2) If the system at some instant is in a particular state Ψ, what are the probabilities of those possible outcomes?

(3) What is the state immediately after the measurement is completed and a particular outcome has been obtained?

For position observables we've already addressed questions 1 and 2 and have found (or rather, postulated) that all locations are allowed, just as classically; and that the probability distribution function is given by Eq. (4.6). For the energy observable, the answer to question 1 is that the allowed energies are the eigenvalues E of Eq. (4.1); namely, those energies for which that equation has well-behaved solutions (eigenfunctions). The generalization to other observables turns out to be as follows. Corresponding to each physical observable there is a distinctive eigenvalue equation, analogous to Eq. (4.1) for energy. Just what these equations are for the various observables of interest is a subject that cannot be easily summarized here. We will have some discussion of this topic later on. For the present, just accept that every physical observable has its own well-defined equation involving an initially undetermined parameter. The values of that parameter for which there are well-behaved solutions are the eigenvalues of the observable in question; the corresponding solutions are the eigenfunctions of that observable. Constantly bear in mind that different physical quantities have different sets of eigenfunctions. In answer to question 1, the assertion is that the possible outcomes of a measurement—the *spectrum* of the observable—is the set of eigenvalues of the equation corresponding to that observable; those and only those outcomes.

If the system happens at some instant to be in an eigenstate of an observable of interest, we further postulate that a measurement of the observable at that instant will uniquely yield the corresponding eigenvalue. Typically, though, the system will not be in an eigenstate of the observable of interest, or indeed in an eigenstate of any physical observable. This then leads us to the generalized question 2 above. For a general state Ψ, the outcome of a measurement on some observable of interest will be distributed in a probabilistic manner. What is the probability distribution? It will be easiest at first to state the postulated answer for the case where the spectrum of the observable is countable, or *discrete* as one says. Let the eigenfunctions u_n be labeled by an index n, and let λ_n be the eigenvalue corresponding to the nth eigenfunction u_n. Assume that the eigenfunctions have been normalized. Supposing that the system is in a state Ψ, the quantum rule is then as follows. Define a *probability amplitude* A_n according to

$$A_n = \langle u_n | \Psi \rangle, \tag{4.7}$$

recalling from Eq. (4.4) the definition of the scalar product. The probability P_n for the outcome λ_n is then asserted to be

$$P_n = A_n^* A_n. \tag{4.8}$$

For observables such as position and momentum that have a continuous spectrum, let u_λ be the eigenfunction corresponding to eigenvalue λ, where the allowed values of λ now lie in a continuum. Given the system's state Ψ, one does not in this situation ask for the probability of finding one particular value of λ, but rather for the probability $P(\lambda)d\lambda$ of finding the observable in an infinitesimal interval $d\lambda$. Just as in the discrete case, define a probability amplitude $A(\lambda)$ according to

$$A(\lambda) = \langle u_\lambda | \Psi \rangle. \tag{4.9}$$

The probability density is then asserted to be

$$P(\lambda) = A(\lambda)^* A(\lambda). \tag{4.10}$$

For an observable whose spectrum is mixed, a portion being discrete and another portion continuous, equations (4.7) and (4.8) apply for the discrete part, (4.9) and (4.10) for the continuum part.

There is a third question to be asked beyond possible outcomes and probability distributions. Immediately after a measurement has been made and a particular result λ_n obtained (for simplicity, we again assume a discrete spectrum), what is now the state of the system? The system has been disturbed by the measurement process, so its wave function just after the measurement is not what it was just before. What is the new wave function? The quantum assertion is as follows. Whatever the state of the system was just before the measurement, during the measurement process it "collapses" into the eigenstate u_n that corresponds to the eigenvalue λ_n obtained in the measurement. The wave function then develops in time under the governance of the Schroedinger equation. It has to be said that there is a great deal of idealization involved in this assertion, since, among other things, measurements are not really accomplished in an instant. In addition, the very notion of measurement itself, which here we are treating as an unanalyzed action performed from the outside on our quantum system, poses deep technical and, ultimately, philosophical problems. But for the present, let's stick with the naive proposition enunciated above.

The principles of quantum mechanics have been illustrated so far mainly on the example of a single particle. The extension to multiparticle systems is straightforward, although in the working out of actual applications the mathematics can become enormously more difficult. For a multiparticle system, the state is described by a wave function depending on time and on as many position vectors as there are particles. Of course, the potential energy will in general also depend on all those position variables. In Eqs. (4.1) and (4.2), there will now be a sum of terms analogous to the first one on the left-hand side of those equations, one such term for each particle, each term with its own mass and with its own position variables in the derivatives. The scalar product defined in Eq. (4.4) is now extended to involve integration over the position variables of *all* the particles. With that understood, Eqs. (4.7)–(4.10) remain unchanged. In contrast to the one-particle interpretation expressed in Eq. (4.6), the product $\Psi^*\Psi$ now gives the joint probability distribution in the location of all the particles. It must also be stressed here that we have yet to deal with spin angular momentum, a dynamical attribute that is possessed by such particles as electrons, protons, and neutrons. We will take up spin later on.

Commuting Observables

For what is to follow it will be useful to introduce the notion of linear independence. A function F is said to be a *linear combination* of a set of n functions $u_1, u_2, \dots, u_n$ if it can be "expanded" in those functions according to

$$F = C_1 u_1 + C_2 u_2 + \cdots + C_n u_n,$$

where the Cs are (possibly complex) constants. The set of functions u_n is said to be a *linearly independent* set if no one of them can be written as a linear combination of the others.

Now consider some particular eigenvalue λ of an observable of interest. It may happen that there is only one linearly independent eigenfunction u_λ corresponding to that eigenvalue. One then speaks of a *nondegenerate* situation. On the other hand, it can happen that there are two or more linearly independent eigenstates all with the same eigenvalue λ. In that case one speaks of *degeneracy*. Often, the occurrence of degeneracy will reflect the fact that the eigenstates of λ are also eigenstates of some other observable; call it μ. In such a situation let us supply a second index on the eigenstate, writing $u_{\lambda,\mu}$. The subscripts tell us that the state in question is *simultaneously* an eigenstate of the one observable, with eigenvalue λ, and of the other observable, with eigenvalue μ. Two observables for which this phenomenon of simultaneous eigenstates occurs are said to *commute*. If the spectrum is discrete, then in the state $u_{\lambda,\mu}$ both observables have definite values; or, as we will say, both are "known." If the spectrum is continuous, then there are states (formed out of narrow superpositions of neighboring eigenstates) in which both can be known to within arbitrary precision. Depending on the particular observables in question, it can happen, even when the eigenvalues λ and μ are both specified, that there is still some remaining degeneracy; that is, that there are two or more states sharing the same λ and μ eigenvalues. In that case there may be yet other observables that commute with both λ and μ. One finally has a *complete* set of commuting observables when the simultaneous eigenstates of all are uniquely specified by the simultaneous eigenvalues.

The three Cartesian position components x, y, z form a set of commuting observables with continuous spectra. One can certainly form wave functions that are as localized as one wishes in all three variables simultaneously. So too for the three components p_x, p_y, p_z of momentum. But there are other sets of observables that do not commute; for example, x and p_x do not commute. For such pairs, there are no states for which both observables can be known with unlimited precision; indeed, the limits are set by Heisenberg's uncertainty principle.

The Uncertainty Principle

Consider some particular observable, for example, the x coordinate of a particle's position. There are wave functions for which the spatial probability distributions are as narrowly peaked as one wishes about some particular value of x. So too for the momentum component p_x. But as Heisenberg was the first to show, there is a limit to the possibility of *simultaneous* peaking in both of these quantities. Given the wave function Ψ, we already know how to find the spatial probability distribution. We have not yet said how to extract from Ψ the momentum distribution. But there are definite rules for this, as we will soon discuss. It turns out that if the spatial distribution is narrow, the momentum distribution will inevitably be broad, and vice versa. There's no getting around it. A measure of the spread of any distribution is the "root mean square deviation" about the average. Illustrated on the example of the position coordinate x, the meaning is as follows. Given the probability distribution, one can compute the average value of x, denoted by $\langle x \rangle$; also the average value of x^2, denoted by $\langle x^2 \rangle$. Now if the distribution were peaked infinitely sharply about one particular x, so that every trial yielded the same value of x, then all values of x^2 would also be the same. It would then be the case that $\langle x^2 \rangle = \langle x \rangle^2$. For all other distributions, one can easily see that $\langle x^2 \rangle$ must be larger than $\langle x \rangle^2$: not much larger if the distribution in x is strongly peaked, much larger if the distribution is broadly spread out. The root mean square deviation is defined by

$$\Delta x \equiv \sqrt{\langle x^2 \rangle - \langle x \rangle^2}.$$

It is a useful measure of the spread in the distribution. Small Δx means a narrow, large Δx a broad distribution.

If we know the state Ψ at a certain instant we can work out the spatial spread Δx and also work out the probability distribution in momentum and hence the root mean square deviation Δp_x. What Heisenberg showed is that for *any* wave function Ψ, the following inequality holds:

$$\Delta x \cdot \Delta p_x \geq \frac{1}{2}\hbar. \tag{4.11}$$

This sets a limit on how well one can know the two observables simultaneously. There are similar bounds for the pairs (y, p_y) and (z, p_z); and there are still other bounds for other pairs of observables that do not commute. But there are no limits to how well one can simultaneously know, for example, x and p_y. These two commute. The uncertainty principle can be expressed in quite general terms for any pair of observables, but we will not write out the general result here, since that would require substantial technical excursions.

There is another inequality relation in quantum mechanics that is often cited, one involving energy and time. It has the look of an uncertainty principle but it stands on a different footing than the Heisenberg uncertainty relations discussed above. Let us inquire about this. Suppose that the system is in a state described by the wave function $\Psi(x, y, z, t = 0)$ at some initial instant $t = 0$. There will be some probability distribution in energy for this state; correspondingly, there will be a root mean square deviation ΔE that measures the spread of that energy distribution. At some later time t, the wave function will, of course, have changed. But for small enough t one expects that the wave function will not have changed much. It may be asked, how much time has to elapse before the wave function first differs appreciably from what it was at the initial time? Call this time τ. "Differs appreciably" is of course not a highly precise phrase. It can be defined more precisely, but here let's just be loose. One finds that the time τ is coupled with the root mean square energy deviation by the inequality relation

$$\tau \cdot \Delta E \geq \hbar. \tag{4.12}$$

This is sometimes referred to as a time-energy uncertainty relation, but that is not a good way to look at it. Time is a dynamical quantity, of course, in that it varies with time! But it does so trivially, self-referentially. It is *the* independent variable on which other things depend: wave functions, probability distributions of various observables, and so on. Time itself just moves relentlessly on—there is no quantum mechanical notion of probability about it (although, of course, for real-life clocks there is plenty of reason to speak in a practical sense of a probability spread concerning the clock's accuracy). Equation (4.12) should just be taken for what it is, as expounded above.

We have said that for every physical observable there is a concrete eigenvalue equation that determines the spectrum and corresponding eigenfunctions. The eigenvalues are obviously of direct physical interest in themselves. The corresponding eigenfunctions are of interest for determining the probabilities of the various outcomes of a measurement of the observable, given the state Ψ of a system [see Eqs. (4.7)–(4.10) and the multiparticle generalizations discussed thereafter]. For the energy observable, we have already written down the eigenvalue equation—Eq. (4.1)—for the case of a single particle; and we have indicated how it generalizes to a system of two or more particles. But what about other observables? In what follows in this chapter, we will focus on the momentum, orbital angular momentum, spin angular momentum, and energy observables of a single particle; and on ancillary matters.

Momentum

The eigenvalue equations for the Cartesian components of momentum $\mathbf{p}$ turn out to be very simple. For example, for the component p_x the equation is

$$-i\hbar\frac{\partial u}{\partial x} = p_x u. \tag{4.13}$$

There are analogous equations for the other two components. The three Cartesian components of momentum commute. That is, one can find solutions that are simultaneous eigenstates of all three. It is easy to check from Eq. (4.13) and its analogs for the other Cartesian components that the unique state with simultaneous eigenvalues p_x, p_y, p_z (collectively, the three-vector $\mathbf{p}$) is

$$u_{\mathbf{p}}(x, y, z) = \left(\frac{1}{2\pi\hbar}\right)^{3/2} \exp(i\mathbf{p}\cdot\mathbf{r}/\hbar),$$
$$\mathbf{p}\cdot\mathbf{r} = xp_x + yp_y + zp_z. \tag{4.14}$$

The numerical coefficient in front of the exponential has been fixed to serve a later purpose. The above solution simultaneously solves Eq. (4.13) and its counterparts. It is evident that momentum is not quantized: all three-vectors $\mathbf{p}$ are allowed. Momentum shares this property with the position observable $\mathbf{r}$, all locations being allowed.

Suppose that our particle is in a state described by some wave function Ψ. What will be the distribution of outcomes of a momentum measurement? According to Eqs. (4.9) and (4.10), the probability amplitude is

$$A_{\mathbf{p}} = \langle u_{\mathbf{p}} | \Psi \rangle; \tag{4.15}$$

where, again, see Eq. (4.4) for the definition of the above scalar product. The momentum probability density is then

$$P(\mathbf{p}) = A_p^* A_{\mathbf{p}} \tag{4.16}$$

with the meaning that P, integrated over some finite region of the momentum variable, is the probability that the momentum will be found to be in that region. As an aside, it may be interesting to give an example here of the best one can do within the constraints of the Heisenberg uncertainty principle. For simplicity, take the case of one-dimensional motion along the x axis. Here is a special wave function that is chosen to be representative of a whole family of states that minimize the position-momentum uncertainty relation:

$$\Psi = N \exp(-x^2/4\lambda^2),$$

where N is an appropriate normalizer that we need not spell out and λ is an arbitrary parameter. The probability distribution function in x is just $P(x) = \Psi^*(x)\Psi(x)$. From this it is easy to work out various averages, in particular the mean square deviation in position. The result is $\Delta x = \lambda$. Using Eqs. (4.15) and (4.16) we can also work out the momentum probability distribution function and hence the mean square deviation in momentum. The result is $\Delta p_x = \hbar/2\lambda$. The product of space and momentum spreads is therefore $\Delta p_x \Delta x = \hbar/2$, which is exactly equal to the smallest possible permitted by Heisenberg; see Eq. (4.11).

The Operator Concept

Where did momentum eigenvalue equation (4.13) come from? A plausible basis for it can be presented along the following lines. We have already agreed to accept the Schroedinger equation (4.2) and the expression (4.6) for the spatial probability distribution. From the latter, if we are given the system wave function Ψ, we can compute the average values (*expectation values*) of various spatial quantities. In particular, consider the average $\langle x \rangle_t$ of the position observable x at time t. From Eq. (4.6) it follows that

$$\text{(i)} \qquad \langle x \rangle_t = \int\int\int dx\, dy\, dz\, \Psi^* x \Psi.$$

This expectation value varies with time because the wave function does. One can work out the time derivative of $\langle x \rangle_t$ using the Schroedinger equation (4.2). The result is

$$m\frac{d\langle x \rangle_t}{dt} = -i\hbar \int\int\int dx\, dy\, dz\, \Psi^* \frac{\partial}{\partial x}\Psi.$$

But classically, the x component of momentum is just $p_x = mdx/dt$. This strongly suggests that the quantum mechanical expectation value of p_x is

$$\text{(ii)} \qquad \langle p_x \rangle_t = \int\int\int dx\, dy\, dz\, \Psi^* \left(-i\hbar\frac{\partial}{\partial x}\right)\Psi.$$

There are similar equations for the other Cartesian components p_y and p_z, where the differentiations are with respect to y and z, respectively.

With the function $A_{\mathbf{p}}$ defined in Eq. (4.15), which is based on $u_{\mathbf{p}}$ as defined in Eq. (4.14), it is now a purely mathematical consequence of (ii) and of Eq. (4.4) that

$$\langle p_x \rangle_t = \int\int\int dp_x\, dp_y\, dp_z\, A^*_{\mathbf{p}} p_x A_{\mathbf{p}};$$

$$\int\int\int dp_x\, dp_y\, dp_z\, A^*_{\mathbf{p}} A_{\mathbf{p}} = 1.$$

These two equations back up our hunch that $A^*_{\mathbf{p}} A_{\mathbf{p}}$ is indeed the probability distribution function for momentum, confirming that Eq. (4.13) has been correctly identified as the momentum eigenvalue equation.

The object within parentheses on the right side of (ii) is what is called the momentum *operator*. Generally, an operator is some rule for acting on a function f to produce what is typically a different function. In this case the rule is, differentiate f with respect to x, then multiply by the factor $(-i\hbar)$. We denote operators with a tilde, so that the three Cartesian components of the momentum operator are

$$\tilde{p}_x = -i\hbar\frac{\partial}{\partial x}, \;\; \tilde{p}_y = -i\hbar\frac{\partial}{\partial y}, \;\; \tilde{p}_z = -i\hbar\frac{\partial}{\partial z}. \qquad (4.17)$$

For example, when $\tilde{p}_x$ acts on any function f, it yields the function g, where $g = -i\hbar\partial f/\partial x$. The momentum eigenvalue equation (4.13) can thus be written $\tilde{p}_x u = p_x u$.

In this way of looking at things, we see what it is that is special about eigenfunctions of the momentum operator. Take the x component of momentum as an example. When the operator acts on an arbitrary function it typically produces a different, linearly independent function. But when it acts on an eigenfunction of momentum, denoted here by u, it gives back that same eigenfunction multiplied by a number. That number is the eigenvalue p_x. This is the general situation. One somehow identifies an operator corresponding to an observable, then sets up the eigenvalue equation. If $\tilde{B}$ is the operator, the form of that equation is $\tilde{B}u = bu$, where b is a parameter. Each value of b for which there is a well-behaved solution u is an eigenvalue; the function u is the associated eigenfunction. And our basic assertion is that the eigenvalues are the allowed outcomes of a measurement of the observable.

The operators corresponding to Cartesian components of momentum we have already discussed. The operators for the components of position are even simpler. For example, the *operator* $\tilde{x}$ acts on any function $f(x, y, z)$ to just multiply it by the variable $x : \tilde{x}f = xf$; and similarly for the other position observables.

We have now identified the operators corresponding to the position and momentum observables. What about other observables? For energy we already have the eigenvalue equation; in the case of a single nonrelativistic particle, Eq. (4.1). Let us look at it from the point of view of operators. Classically, the sum of kinetic and potential energies is the total energy E:

$$\frac{1}{2m}(p_x^2 + p_y^2 + p_z^2) + V(x, y, z) = E.$$

To get the quantum mechanical operator for energy, it seems natural to simply replace the classical momenta in the above expression by the corresponding quantum mechanical operators. So too for the potential energy. The potential energy operator acting on a function just multiplies that function by $V(x, y, z)$.

The momentum operators involve differentiation, as discussed before. The energy operator thus obtained is called the Hamiltonian and is marked with a tilde. The energy eigenvalue equation is therefore

$$\tilde{H}u = Eu, \tag{4.18}$$

where

$$\tilde{H} = -\frac{\hbar^2}{2m}\left\{\frac{\partial^2}{\partial x^2} + \frac{\partial^2}{\partial y^2} + \frac{\partial^2}{\partial z^2}\right\} + V.$$

Here u is an eigenfunction and E is the corresponding eigenvalue. We have recovered Eq. (4.1)! We can now also see that the Hamiltonian operator plays a special role in quantum mechanics. It governs the time evolution of the system wave function. The Schroedinger equation (4.2), expressed compactly in terms of the Hamiltonian operator, is

$$\tilde{H}\Psi = i\hbar\frac{\partial\Psi}{\partial t}. \tag{4.19}$$

Other than introducing compact notation, what have we accomplished with the concept of operators? So far, only a kind of consistency. Having identified the position and momentum operators, we could check that the energy eigenvalue equation (4.1) that we started out with is indeed the eigenvalue equation associated with the Hamiltonian operator; and that the latter is obtained from the classical expression for energy by replacing the momentum and position variables there by their quantum operators. Equation (4.2) holds quite generally, whether for a single particle, a collection of particles, or a quantum field system.

This emboldens us to extend the principle to other observables, at least those that have classical incarnations. The procedure is as follows. Take the observable of interest as expressed classically in terms of the position and momentum variables, then obtain the corresponding quantum operator by replacing the position and momentum variables by their quantum operators. We will shortly illustrate this procedure for obtaining the orbital angular momentum operator.

Commutation Relations

Given any two operators $\tilde{A}$ and $\tilde{B}$, and a function f, the expression $\tilde{A}\tilde{B}f$ stands for the function that results when f is first acted on by $\tilde{B}$ and the result then acted on by $\tilde{A}$. The order of operators can matter. That is, it can happen that $\tilde{A}\tilde{B}f \neq \tilde{B}\tilde{A}f$. The operator product difference $\tilde{A}\tilde{B} - \tilde{B}A$ is called the commutator of the two operators.

Here is one example of an operator commutation relation. Consider the position and momentum operators that we have identified above. It is easy to check that for an arbitrary function f,

$$\tilde{x}\tilde{p}_x f = -i\hbar x \frac{\partial f}{\partial x}; \quad \tilde{p}_x \tilde{x} f = -i\hbar \frac{\partial}{\partial x}(xf) = -i\hbar x \frac{\partial f}{\partial x} + i\hbar f.$$

Since this holds for arbitrary functions f, it implies for the operators the *commutation relation*

$$\tilde{x}\tilde{p}_x - \tilde{p}_x \tilde{x} = i\hbar. \tag{4.20}$$

A commutation relation is an equation involving the difference of the product of two operators taken in opposite orders. If the two orders give the same result, the operators are said to *commute*. It is easy enough to check the commutation relations among the other components of the position and momentum observbles. Thus, one finds that $\tilde{x}$ and $\tilde{p}_y$ commute, as do $\tilde{y}$ and $\tilde{p}_x$, and so on.

Some final words about operators. In the abstract formulation of quantum mechanics the operator concept plays a central role. In the approach that we have been taking, the state of a system at any instant is described concretely by a function of spatial coordinates; and the operators we have met involve concrete operations such as differentiation. In the abstract formulation, the possible states form a mathematical space of abstract objects called "vectors," and operators are rules for mapping abstract vectors into generally different vectors in that space. This elevated vantage point affords great flexibility and breadth of view. However, for practical results it is often best to descend to some concrete *representation*. That's what we have been doing from the start, working in the so-called Schroedinger "position space" representation.

Angular Momentum

Orbital

The angular momentum of a particle is defined classically in terms of the position and momentum observables by the cross-product relation $\mathbf{L} = \mathbf{r} \times \mathbf{p}$. In terms of Cartesian coordinates: $L_x = yp_z - zp_y$, $L_y = zp_x - xp_z$, $L_z = xp_y - yp_x$. In addition to the three Cartesian components of **L**, we will also want to consider the magnitude of the angular momentum; or more conveniently, its square L^2. The corresponding quantum mechanical operators follow from the principle adopted in the discussion immediately following Eq. (4.19); namely, take the classical expressions and replace the position and momentum variables by their quantum operators. For example, the quantum operator corresponding to the z component of angular momentum is

$$\tilde{L}_z = -i\hbar\left\{x\frac{\partial}{\partial y} - y\frac{\partial}{\partial x}\right\}.$$

The eigenvalue problem for this component is therefore expressed by the differential equation

$$-i\hbar\left\{x\frac{\partial u}{\partial y} - y\frac{\partial u}{\partial x}\right\} = L_z u.$$

Here L_z on the right is the eigenvalue. There are similar expressions for the other components and for the square of the angular momentum. It actually turns out best to express the operators for the *Cartesian* components in terms of *spherical coordinates* r, θ, ϕ.

Angular momentum is much more interesting in the quantum world than in the classical. It displays a number of features that have a distinctly odd flavor. Unlike the position and momentum observables in terms of which it is defined, it can take on only certain discrete values; and there are oddities beyond that. One important quantum mechanical property of angular momentum is the following. With an isolated exception, there are no simultaneous eigenstates of all three components, or indeed of any pair of components of **L**. That is, the components of the angular momentum vector do not commute among themselves; thus there are no states for which one "knows" the values of any pair of angular momentum components, states for which there is a definite outcome for each of the components. However, each Cartesian component *does* commute with L^2. Hence there *do* exist simultaneous eigenstates of L^2 and the component of **L** in *any* direction, not even necessarily along a coordinate axis. For definiteness, we will focus on the simultaneous eigenstates of L^2 and L_z. The eigenvalue problem for L_z has been written out above. It simplifies when expressed in spherical coordinates. That for L^2 is complicated even in spherical coordinates. We will not write down the eigenvalue equations but merely quote some results. For the observable L^2 the eigenvalues—the allowed values—are given by

$$L^2 = \ell(\ell+1)\hbar^2, \quad \ell = 0, 1, 2, 3, \ldots, \tag{4.21}$$

where ℓ ranges from zero to infinity in integer steps. For given "quantum number" ℓ, the eigenvalues of L_z are

$$L_z = m_\ell \hbar, \quad m_\ell = -\ell, -\ell+1, \ldots, \ell-1, \ell. \tag{4.22}$$

Thus, for given ℓ the quantum number m_ℓ ranges in unit steps over $2\ell+1$ possible values from $-\ell$ to $+\ell$.

In the eyes of a classicist, there are several strange effects on display here. For one thing, the magnitude of the angular momentum vector is quantized: it can take on only certain discrete values. This is so even though the angular momentum operator is defined in terms of position and momentum operators that have continuous spectra. Moreover, for a given one of the allowed values of L^2—that is, for given quantum number ℓ—we see from Eq. (4.22) that the projection of **L** onto the z axis can take on only certain discrete values. So there is some sort of additional quantization going on here. Can it be that **L** points only in certain discrete directions in space? If so, the z axis cannot be one of those directions. After all, if **L** pointed exactly along the positive or negative z direction, the square of that projection onto the z axis would have to be equal to L^2, in which case we would expect $L_z^2 = L^2 = \ell(\ell+1)\hbar^2$. But from Eqs. (4.17) and (4.18) we see (for given ℓ) that the biggest possible value of L_z^2 is $\ell^2\hbar^2$, which is smaller than $\ell(\ell+1)\hbar^2$. Since there is nothing special about how we choose to orient our coordinate axes, we might reorient so that one of the putatively allowed directions of **L** is taken as the new z axis. But then the above argument would again tell us that **L** cannot point along the direction in which it is said to point! The way out of these absurdities is to abandon classical images of the angular momentum vector as pointing in any definite directions in space. Quantum mechanics is odd!

But classical thinking isn't bad for macroscopic situations. The microscopic unit of angular momentum is $\hbar$. That is a very tiny quantity on the scale of what is encountered in the everyday world. A jelly bean spinning at all noticeably has an angular momentum magnitude that is hugely large compared to $\hbar$. Here, where very large ℓ values are in play, the fractional change in L^2 when one goes from ℓ to $\ell+1$ is very tiny. In the macroscopic range, therefore, the allowed values of L^2 practically form a continuum, just as classically. So too the illegal notion of definite directions of **L** becomes almost legal for physically realistic, macroscopic situations.

Let us return to the angular momentum eigenvalue problem and focus on the eigenfunctions; call them u_{ℓ,m_ℓ}. They now bear the two quantum numbers indicated. The eigenfunctions are best expressed in terms of spherical coordinates r, θ, ϕ (the "polar" angle θ is the angle between the position vector **r** and the z axis; the "azimuthal" angle ϕ is the angle between the x axis and the projection of **r** onto the x-y plane). It turns out that each function u_{ℓ,m_ℓ} is the product of a definite function of the angles multiplied by a function of the radial variable r:

$$u_{\ell,m_\ell} = R(r)Y_\ell^{m_\ell}(\theta, \phi).$$

The function $R(r)$ is arbitrary so far as angular momentum is concerned. However, the *spherical harmonics* $Y_\ell^{m_\ell}$ are definite functions of the angular variables. Here are a few of them, just for show:

$$Y_0^0 = \sqrt{\frac{1}{4\pi}}, \quad Y_1^1 = \sqrt{\frac{3}{8\pi}} \sin\theta e^{i\phi},$$

$$Y_1^0 = \sqrt{\frac{3}{4\pi}} \cos\theta, \quad Y_1^{-1} = \sqrt{\frac{3}{8\pi}} \sin\theta e^{-i\phi}.$$

Spin

For certain species of particles there is an intrinsic kind of angular momentum, *spin*, that is in addition to the angular momentum associated with orbital motion. Among the particles that have this property are the constituents of ordinary matter, electrons, protons, and neutrons. As discussed briefly in Chapter 2, in thinking about spin one is tempted to picture the particle as a tiny sphere, the spin angular momentum arising from presumed rotation about an axis through the center of the particle. The earth's motion suggests an analogy. The earth has orbital angular momentum associated with its motion around the sun, but also spin angular momentum arising from its rotation about the polar axis. But this picture has its limits in the world of microscopic particles. Quantum mechanically, what is germane is simply that for certain kinds of particles there is a vector observable $\mathbf{S}$ which is in addition to the observables defined in terms of position and momentum; and the Cartesian components of $\mathbf{S}$ relate to one another in the same way as do the components of orbital angular momentum $\mathbf{L}$. The components of $\mathbf{S}$ do not commute with one another, but every one of them commutes with S^2.

What is distinctive about spin angular momentum, what sets it apart from the orbital kind, is that the magnitude is not a dynamical variable at all. In the orbital case, the possible outcomes of a measurement of L^2 are the eigenvalues given in Eq. (4.21). One has the quantum mechanical oddity that the spectrum is not continuous, as it would be classically, but at least there are infinitely many possible outcomes. For spin, S^2 is a fixed quantity characteristic of the particle species. Its value is

$$S^2 = s(s+1)\hbar^2, \tag{4.23}$$

where, depending on the species, s is some definite integer or half-odd integer. For any of the Cartesian components, say S_z, there are the $2s+1$ eigenvalues

$$S_z = m_s\hbar, \quad m_s = -s, -s+1, \ldots, s-1, s. \tag{4.24}$$

The above two equations have the look of Eqs. (4.21) and (4.22); but as said above, unlike the orbital quantum number ℓ, s does not take on a range of possible values. It is fixed! There is another contrast with orbital angular momentum. For the latter, the quantum number ℓ is necessarily restricted to integers. For spin, the parameter s may be half-odd integral or integral. Those are the only two possibilities allowed by general quantum mechanical considerations. For electrons, protons, and neutrons it happens that $s = 1/2$; for pions, $s = 0$; and so on for the other particles of nature. The difference between half-odd integral and integral spin is not a small, technical matter. The distinction is profound. More on that later, except to say here that the world would be a very different place and we would not be in it if the electron, proton, and neutron were particles of integer spin.

Returning to Eq. (4.24), we see that there are $2s+1$ spin degrees of freedom; that is, there are that many linearly independent eigenstates of S_z. For the electron and other spin-one-half particles, therefore, there are are only two spin degrees of freedom. A general spin state is a linear combination of the two eigenstates of S_z. There is of course nothing magic about the z axis. The possible outcomes of a measurement of spin angular momentum along *any* direction can only be $\pm\hbar/2$. The eigenstate of the component of angular momentum in some one direction is not an eigenstate of the component in another direction. This is so whether for the spin or orbital angular momentum. Here is an illustration based on the spin of an electron (or any other $s = 1/2$ particle). Suppose that the electron is in an eigenstate of S_x with eigenvalue $+\hbar/2$. For that state, a measurement of the x component of spin must yield that outcome with 100% probability. This same state is, however, a linear combination of the eigenstates of S_z. A measurement of the z component of spin has the possible outcomes $\pm\hbar/2$; both, as it happens in this particular example, with equal probability.

Total Angular Momentum

A particle with spin has two kinds of angular momentum, orbital **L** and spin **S**. It is natural to define a total angular momentum observable by

$$\mathbf{J} = \mathbf{L} + \mathbf{S}. \tag{4.25}$$

It turns out that the Cartesian components of **J** relate to one another just as do the components of **L** or **S** among themselves. As in those other cases, the Cartesian components of **J** do not commute with one another, but the projection of **J** along *any* direction commutes with the squared angular momentum J^2. We will again single out the component J_z. Moreover, in what follows we will restrict ourselves to the simple and highly relevant case $s = 1/2$. The eigenvalues of J^2 are

$$J^2 = j(j+1)\hbar^2, \quad j = 1/2, 3/2, 5/2, \ldots; \tag{4.26}$$

and, for given j, the eigenvalues of J_z are the $2j+1$ quantities

$$J_z = m_j\hbar, \quad m_j = -j, -j+1, \ldots, j-1, j. \tag{4.27}$$

There is now something else. The observables J^2, J_z, and L^2 all commute with one another, so there are simultaneous eigenstates not only of J^2 and J_z but also of these, and L^2. The simultaneous eigenstates of all three observables bear the three quantum numbers j, m_j, and ℓ. It may be asked: for given j what are the possible values of ℓ? The answer is that there are only two values:

$$\ell = j + 1/2, \quad j - 1/2. \tag{4.28}$$

Aspects of Energy

Much of the workaday effort of practitioners of quantum mechanics is devoted to the energy observable—confronting the energy eigenvalue problem, looking for physically reasonable approximations where exact solutions are beyond reach (as is most often the case), and developing physical intuition. The momentum and angular momentum eigenvalue problems can be solved exactly, and once solved remain solved. But the energy situation varies from one physical problem to another, depending on the details of the potential energy function. The energy observable is of interest for another reason as well, one that gives it special status among all quantum observables. The Hamiltonian, which is the operator corresponding to energy, governs the time evolution of a physical system in the sense reflected in Eq. (4.19). Although we have been illustrating the principles of quantum mechanics for a one-particle system, that equation holds for multiparticle systems as well, with the Hamiltonian extended in the manner described earlier.

Time Evolution

The time evolution problem is that of finding the wave function at general time t if it is known at some initial time. To this end, suppose that we can solve the energy eigenvalue problem, so that we have the full set of linearly independent energy eigenfunctions u_n and their corresponding energy eigenvalues E_n. It is a mathematical fact, essential to the whole interpretative apparatus of quantum mechanics, that the collection of all the eigenstates of *any* physical observable forms a *complete* set. What is meant by this is that any well-behaved function can be expressed as a linear combination of the eigenfunctions. In particular, the actual system wave function $\Psi(t)$ at time t can be expanded in the energy eigenfunctions u_n:

$$\Psi(t) = A_1(t)u_1 + A_2(t)u_2 + A_3(t)u_3 + \cdots, \qquad (4.29)$$

where the coefficients $A_n(t)$ carry the time variation, the eigenfunctions depending on space but not time. The wave function

Ψ and the functions u_n all depend on the spatial variables, but we are not displaying those variables here. Suppose that we know the function $\Psi(0)$ in its dependence on spatial variables at some initial time $t = 0$. We then know the expansion coefficients $A_n(0)$ at that initial time. But one can readily show from Eqs. (4.18) and (4.19) that the coefficient $A_n(t)$ at general time t is related to its value at time $t = 0$ by the simple equation

$$A_n(t) = A_n(0)\exp(-iE_n t/\hbar). \tag{4.30}$$

The problem of time evolution of the system wave function is hereby solved—insofar as the energy eigenvalue problem can be solved. Of course, this victory may look to be hollow since the sum in Eq. (4.29) typically contains an infinite number of terms. But this formal solution provides many insights and serves as a basis for various approximation procedures.

It's interesting to have a look at time evolution for the simplest of all situations, a particle moving freely, $V = 0$. To simplify further, take the case of one-dimensional motion. Classically, if the particle starts out at $t = 0$ with initial position x_0 and initial momentum p_0, the momentum at later time t remains unchanged and the position changes according to $x(t) = x_0 + p_0 t/m$. Quantum mechanically, we deal with probability distributions. Let $\langle x\rangle_t$ and $\langle p\rangle_t$ be, respectively, the mean position and mean momentum at time t. Similarly, introduce $\langle x^2\rangle_t$ and $\langle p^2\rangle_t$, the respective averages of the square of position and square of momentum at time t. The quantum analog of the classical constancy in momentum turns out to be that the momentum *distribution* does not change with time for a free particle. Therefore, $\langle p\rangle_t = \langle p\rangle_0$ and $\langle p^2\rangle_t = \langle p^2\rangle_0$. But the mean position does change with time. It does so in the same way, in terms of averages, as for the classical position:

$$\langle x\rangle_t = \langle x\rangle_0 + \langle p\rangle_0 t/m.$$

What is more interesting is the mean square deviation in position, a concept that does not arise in the classical case. The mean square deviation is a measure of the spread of the probability distribution. One often speaks of this distribution as describing a *wave packet*, which might be pictured as a disturbance moving bodily through space but also changing its shape as time goes on. We define the mean square spreads in position and momentum by

$$\langle \Delta x^2 \rangle_t = (\langle x^2 \rangle_t) - (\langle x \rangle_t)^2; \quad \langle \Delta p^2 \rangle_t = (\langle p^2 \rangle_t) - (\langle p \rangle_t)^2.$$

It is easy to show that the mean square spread in position varies with time according to

$$\langle \Delta x^2 \rangle_t = \langle \Delta x^2 \rangle_0 + bt + \langle \Delta p^2 \rangle_0 t^2/m^2.$$

The coefficient b in the term linear in time depends on further details of the initial wave function and is not of special interest here. What is interesting is the last term. The coefficient of t^2 is necessarily positive. Thus, whatever may be the sign of b, after a long enough time the packet not only moves but widens. That is, however localized the packet may have been at some initial time, eventually, as time goes on, the particle is increasingly "spread out."

Tunneling

Suppose that a particle moves one-dimensionally in the potential $V(x)$ shown in Fig. 4.2. The rather complicated, wiggly potential function shown there is chosen to help illustrate certain interesting features of the energy eigenvalue problem. We will want to contrast the classical and quantum approaches.

Classical Barriers

Classically, the kinetic and potential energies of a particle change as it moves about in its orbit, but the sum $E = K + V$ is a constant of the motion. Since the kinetic energy $K = p^2/2m$ is necessarily non-negative, a classical particle of energy E can move only in regions of space where $V(x) \leq E$. On the energy diagram, the potential varies with x but the total energy E, since it is a constant of the motion and thus independent of x, is represented by a straight, horizontal line. The first thing to

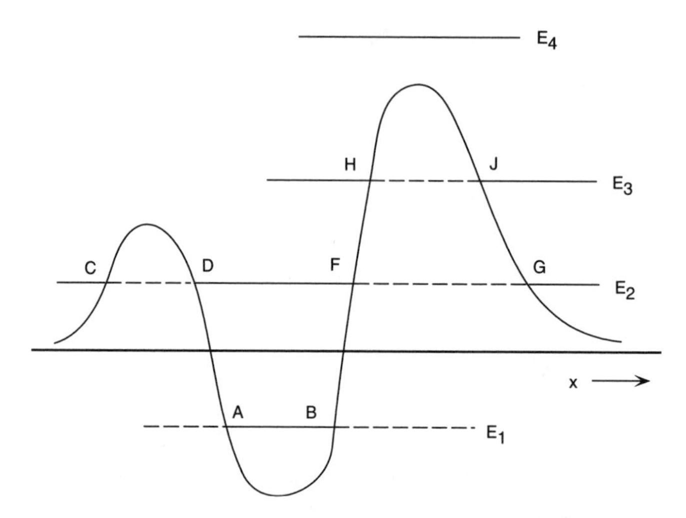

Figure 4.2 An imagined potential function $V(x)$, devised for pedagogical purposes (bold curve). The horizontal lines E_1 to E_4 correspond to various total energies, kinetic plus potential.

be said is that all energies E above the minimum of the potential are classically possible. What the energy actually happens to be is a matter of initial conditions. Let us then consider several different choices of energy E.

(1) With the energy $E_1 < 0$ shown in the diagram, the particle can move only in the finite region between the "turning points" at A and B. It is said to move in a *bound* orbit. If at some moment the particle is moving to the right, it will eventually come to momentary rest when it reaches B. It will then turn around and move to A; then turn back again; and on and on, back and forth between the turning points. The line marked E_1 is drawn solid in the allowed region, dotted in the classically forbidden region.

(2) At the energy E_2, as shown, there are three disconnected orbital regions. One is an *unbounded* region between negative infinity and the turning point at C. Another is the unbounded region between positive infinity and the turning point at G. A third is the bound orbit between the turning points at D and

F. If at the some initial instant the particle gets started to the left of C but moving to the right, it will reach the turning point at C, then turn around and move off to negative infinity; if initially moving to the left, it will head straight to negative infinity. Similar remarks hold for a particle that starts to the right of G; it ends up moving toward plus infinity, either directly or after turning around at G. The orbit between D and F is a bound orbit, as in case 1 above. A particle in such an orbit has enough energy to escape to plus or minus infinity, but it cannot get through the intervening barriers. Similarly, though energy conservation would permit it, a particle cannot get from one of the unbound regions to the other. There are barriers in between.

(3) At the energy E_3, as shown, there are two unbounded orbital regions, one with turning point at H, the other with turning point at J. They do not connect. There is a barrier in between.

(4) At any energy E_4 above the maximum of the potential, there is a single orbital region extending from minus to plus infinity. There are no turning points. If a particle moves in from the far left it will just keep on moving toward plus infinity; conversely, if it comes in from the right it will move to negative infinity. There are no turnings back.

The Quantum Mechanical Case

The first thing to be said is that, despite all its oddities, quantum mechanics shares with classical mechanics the property that the energy E can never lie below V_{min}, the minimum of the potential energy. But whereas all energies $E > V_{min}$ are allowed classically, quantum mechanically, depending on the potential, the spectrum may be discrete, continuous, or mixed. We will restrict ourselves in the present commentary to two broad classes of potential.

(1) Potentials that increase to positive infinity as x goes to positive or negative infinity: $V(x) \rightarrow +\infty$ as $|x| \rightarrow \infty$. Classically, all orbits in any such potential are bounded. Quantum

mechanically, the energy spectrum will be discrete (quantized), meaning that the energy eigenvalues are finitely separated.

(2) Potentials that vanish as x goes toward positive and negative infinity: $V(x) \rightarrow 0$ as $|x| \rightarrow \infty$. The potential shown in Fig. 4.2 belongs to this class. In this case the spectrum is continuous for all energies $E > 0$. If the minimum of the potential is positive, $V_{\text{min}} > 0$, that's the end of the matter; there are no eigenvalues for $E < V_{\text{min}}$ and hence none for $E < 0$. If V_{min} is negative over some regions of x, there may or may not be eigenvalues in the range $V_{\text{min}} < E < 0$. If there are, they form a discrete spectrum.

So much for general comments. In order to bring out some additional points let's now turn to the particular potential shown in Fig. 4.2. We will again consider several different regions of energy.

Suppose for $E < 0$ that there *is* at least one bound state, perhaps more. Let E_1 be a bound state energy eigenvalue. The eigenfunction will generally be concentrated in the classically allowed region between the classical turning points at A and B. But that function will also extend out into the forbidden regions to the left of A and right of B. That is, there will be a finite probability of finding the particle in a classically forbidden region! That is the main point here; the particle can penetrate into classically forbidden places.

For all $E > 0$ the spectrum is continuous, but there are some odd quantum mechanical features here too. We turn to that next. Suppose that at some initial time one forms a state which is a superposition of energy eigenstates with energies narrowly spread around the energy E_2 shown in the diagram. The probability distribution associated with this wave function—the *packet*—will move bodily and change shape as a function of time. Arrange it so that the packet starts out far to the left of C and is moving to the right. By construction, the packet represents a particle of almost definite energy E_2. Classically, such a particle would bang up against the turning point at C and turn around. Quantum mechanically, as the packet approaches and starts to "feel" the potential it will begin to split, one part eventually being reflected back toward minus infinity, another moving on past G toward plus infinity. There is thus a finite probability of tunneling—transmission through a classical barrier. Indeed, there are two such barriers to be gone through at the energy being discussed here. There is yet another interesting quantum mechanical feature to be noted. Suppose the packet is initially concentrated in the classical trapped orbit region between D and F. Classically, the particle would, of course, remain trapped in that region. Quantum mechanically, the packet will leak out over time, some part moving on toward plus infinity, the other part toward minus infinity. This is a kind of radioactive decay process.

At the energy E_3 shown in the diagram, there are the same phenomena of transmission and reflection as for the energy E_2 discussed above, now with only a single barrier to be tunneled through.

At energy E_4, which is representative of any energy $E > V_{max}$, the wave packet will encounter no barriers. A classical particle coming in from the far left would sail right on toward plus infinity; and conversely for a particle coming in from the far right. Quantum mechanically, however, there will be reflection as well as transmission, even though there is no barrier. That is, a packet coming in from the far left begins to split as it approaches the region where the packet feels the presence of the potential, one part of the packet eventually sailing through to the far right, another reflecting back toward the far left; and similarly for a packet arriving from the far right.

A word here on terminology. The eigenstates corresponding to a discrete spectrum (or the discrete portion of a spectrum that is mixed) are often spoken of as *bound states*; and one often speaks of *energy levels* instead of *energy eigenvalues*. For the continuous spectrum the question of which energies are allowed is pointless since they are all allowed over the continuous span of the spectrum. Rather, for any given energy one is interested in the information conveyed by the eigenfunction about transmission and reflection phenomena. In three dimensions, this latter generalizes to the phenomenon of *scattering*. A beam of particles of given energy enters into a force field characterized by some potential. The particles scatter in various directions. What are the scattering probabilities as a function of energy and scattering angle? We will turn to these matters later in the broader context of particle collision reactions.

CHAPTER FIVE

Some Quantum Classics

The title of this chapter indicates that we will be paying brief calls on a number of relatively simple problems that either are important in their own right or serve well to illustrate the workings of quantum theory. In all cases in this chapter we will be concerned with a single, nonrelativistic particle of mass m.

The Free Particle

Suppose that the particle is not acted upon by any force at all. In that case, the potential V is a constant and we can take it to be zero. Since the energy is purely kinetic and therefore proportional to the square of the momentum, energy and momentum obviously commute. So let us first look at the momentum eigenvalue problem, initially for the one-dimensional case. The eigenstate u_p corresponding to the momentum eigenvalue p is at the same time an eigenstate of the free Hamiltonian, with eigenvalue $p^2/2m$. According to Eq. (4.13), the momentum function, up to a multiplicative constant that is of no interest for the present discussion, is

$$u_p(x) = \exp(ipx/\hbar).$$

One can check directly that this is a solution of the energy eigenvalue equation with the energy eigenvalue indicated above; namely,

$$-(\hbar^2/2m)d^2u_p/dx^2 = Eu_p, \qquad E = p^2/2m.$$

But notice, there is two-fold energy degeneracy. The energy E determines only the *magnitude* of the momentum; the sign of p can be positive or negative. We can pull all of this together as follows. For the free particle, all non-negative energies are allowed. For given positive energy E, there are two linearly independent eigenstates, which we can take to be $\exp(ikx)$ and $\exp(-ikx)$. Here k is a *positive* quantity defined by

$$k = \sqrt{2mE/\hbar^2}. \tag{5.1}$$

The *general* solution of the energy eigenvalue equation for energy E is the linear combination

$$u_E = A\exp(ikx) + B\exp(-ikx). \tag{5.2}$$

If $B = 0$, the solution describes a particle not only of definite energy E but of definite, positive momentum $p = \hbar k$. If $A = 0$ the solution describes a particle of negative momentum $p = -\hbar k$. The general energy eigenstate is a linear combination of these two momentum eigenstates. A momentum measurement would have two possible outcomes, right moving and left moving, with relative probabilities in the ratio A^*A/B^*B.

For a free particle in three dimensions, it is again the case that energy and momentum commute, but now momentum is a three-vector. The momentum eigenstate corresponding to the vector eigenvalue **p** is given by the exponential function of Eq. (4.14). This function is also an eigenstate of energy, with eigenvalue $E = p^2/2m$. Here the symbol p without boldface represents the magnitude of the vector **p**. Since the energy depends only on this magnitude there is an infinite degree of degeneracy—the vector **p** can point in *any* direction. For given energy E (hence given magnitude p) the general energy eigenstate is a superposition of the exponentials of Eq. (4.14), taken over all directions of **p**.

Particle in a Box

One Dimension

We may take a one-dimensional box to be a region along the x axis bounded by infinite walls at either end. The kind of wall we have in mind is an idealized, impenetrable one, a place where the potential abruptly rises to plus infinity. This infinite jump in potential corresponds to an infinitely strong, repulsive force at the wall. Despite the other oddities of quantum mechanics, this serves to contain the particle just as it does classically. Even a quantum particle cannot tunnel through an infinite wall. The wall imposes a boundary condition on the wave function; it must vanish at the wall. Suppose then that there is a wall at $x = 0$ and another at $x = L$. In between, suppose that the particle moves force free ($V = 0$). Inside the box the general solution of the energy eigenvalue equation is exactly as in Eq. (5.2). As far as the mathematics is concerned, the (complex) constants A and B are arbitrary. But now we have to impose the boundary conditions. To meet the requirement at $x = 0$, we must set $B = -A$. We then observe that the difference of the two exponentials in Eq. (5.2) is proportional to the trigonometric sine function. Therefore, with C a new arbitrary constant,

$$u_E(x) = C \sin kx.$$

But the solution must also vanish at $x = L$, which requires that $\sin kL = 0$. It is well known that the sine function vanishes when its argument is an integral multiple of π. Therefore, the allowed values of k are $k_n = n\pi/L$, $n = 1, 2, \ldots, \infty$. Correspondingly, we see that the energy eigenvalues and corresponding eigenfunctions (all now labeled with an integer subscript n) are

$$E_n = \frac{\hbar^2\pi^2}{2mL^2}n^2, \quad u_n = \sqrt{\frac{2}{L}}\sin\left(\frac{n\pi}{L}x\right), \quad n = 1, 2, 3, \ldots \tag{5.3}$$

The coefficient in front of the sine function has been chosen to normalize the eigenfunction. The eigenstates and energies are labeled by the integer n, which ranges from one to infinity. There is a countable infinity of bound states. Notice for the box that the energies grow without limit as the integer n gets larger and larger. The absolute value of the spacing between one level and its neighbor, $\Delta E_n = E_{n+1} - E_n$, also grows with n. But the *fractional* change gets smaller with increasing n. For large n, the fractional change is given approximately by $\Delta E_n / E_n = 2/n$, which becomes small for large n. In this sense, for macroscopic energies (very large n) it is almost as if the spectrum were continuous.

This simple problem illustrates how the "good behavior" requirement can lead to the quantization of eigenvalues. Here the requirement was that the wave function must vanish at the walls. More typically, in the absence of walls, good behavior amounts to the requirement that the wave function must be bounded, that it not grow infinitely large as $|x|$ approaches infinity.

Three Dimensions

Now take a three-dimensional box, a cube of side L with one corner at the origin $(x, y, z) = (0, 0, 0)$. Again, suppose that the particle moves force free inside the box. We now demand that the wave function vanish at all six walls. The problem is just as easily solved as in the one-dimensional case. The eigenvalues and eigenfunctions are labeled by three non-negative integers, n_1, n_2, n_3. One finds

$$E_{n_1,n_2,n_3} = \frac{\hbar^2 \pi^2}{2mL^2}(n_1^2 + n_2^2 + n_3^2),$$

$$u_{n_1,n_2,n_3} = \sqrt{\frac{8}{L^3}} \sin\left(\frac{n_1 \pi}{L} x\right) \cdot \sin\left(\frac{n_2 \pi}{L} y\right) \cdot \sin\left(\frac{n_3 \pi}{L} z\right). \tag{5.4}$$

We will have some use for this result later on.

The Harmonic Oscillator

The harmonic oscillator appears in various guises in many branches of science. It ranks among the great classics of science for this reason as well as for its pedagogical value.

One Dimension

The oscillator potential corresponds to the force law $F(x) = -Kx$, where K is a positive parameter, the "spring constant." In good approximation, this law describes the restoring force that acts on a mass attached to a real spring when the spring is stretched (or, for $x < 0$, compressed) through a distance x. The potential energy is $V(x) = Kx^2/2$. Notice that the potential grows unlimitedly as the magnitude of x grows larger and larger. We can therefore anticipate that the quantum mechanical energy spectrum will be purely discrete. It will be convenient here to replace the parameter K by a frequency parameter ω *defined* by $\omega = (K/m)^{1/2}$, where m is the mass of the particle. The potential can then be written

$$V(x) = \frac{1}{2}m\omega^2 x^2. \tag{5.5}$$

If the particle has net energy E, its classical motion lies between turning points at $x = x_0$ and $-x_0$, where $x_0 = (2E/m\omega^2)^{1/2}$. The general solution of the classical equation of motion is

$$x(t) = x_0 \sin[\omega(t - t_0)],$$

where x_0 and t_0 are arbitrary parameters so far as Newton is concerned. They are determined by initial conditions. Since the sine function ranges from -1 to $+1$, this solution confirms that the particle moves between turning points at $x = x_0$ and $x = -x_0$, where the parameter x_0 is determined by the energy E in the manner indicated above. The parameter t_0 is a time when the particle is passing through the origin (in the positive direction). The main thing to notice about the solution is that the motion is oscillatory, with angular frequency ω.

Quantum mechanically, after some rearrangements the energy eigenvalue equation reads as follows:

$$\frac{d^2u}{dx^2} + \frac{2mE}{\hbar^2}u - \left(\frac{m\omega}{\hbar}\right)^2 x^2 u = 0. \tag{5.6}$$

As usual, the equation has solutions for *any* value of the energy E, but typically the solutions "blow up"; they grow without limit as x goes toward infinity in one direction or the other. But for certain energies E_n there is a well-behaved solution, the corresponding eigenfunction u_n. It is very well behaved indeed, falling off to zero very rapidly as the magnitude of x becomes large. The allowed energy eigenvalues are given by the well-known and remarkably simple formula

$$E_n = \hbar\omega\left(n + \frac{1}{2}\right), \qquad n = 0, 1, 2, 3, \ldots \tag{5.7}$$

Notice that the ground level has the finite energy $E_0 = \hbar\omega/2$. We will write out only the ground state eigenfunction u_0. It is

$$u_0 = N \exp\left(-\frac{x^2}{2x_0^2}\right), \qquad x_0 \equiv \sqrt{\frac{\hbar}{m\omega}}, \tag{5.8}$$

where N is a normalizing constant that we do not bother to specify here. Although working forward from the differential equation (5.6) to its solution involves some advanced mathematics, *confirming* that the above expression for u_0 *is* a solution for $E = E_0 = \hbar\omega/2$ involves only a bit of differentiating. Please check! Interestingly, the parameter x_0 here is just exactly the classical turning point that corresponds to the energy E_0. Notice also that the wave function begins to fall off rapidly beyond the classical turning points. Nevertheless, there is a substantial probability of finding the particle in the classically forbidden regions $|x| > x_0$.

Three Dimensions

The "spherical" harmonic oscillator potential is

$$V(r) = \frac{1}{2} m\omega^2 r^2, \tag{5.9}$$

corresponding to an attractive radial force $F = -Kr$; we have again defined ω in terms of the spring constant K according to $\omega = (K/m)^{1/2}$. Because the potential "separates" into a sum of terms, each depending on a different one of the Cartesian coordinates—that is, because $r^2 = x^2 + y^2 + z^2$—the solution of the three-dimensional quantum problem reduces to the solution of the one-dimensional problem that we dealt with above. The eigenvalues are sums of three one-dimensional energies; the eigenfunctions are products of the corresponding one-dimensional eigenfunctions. Let u_n be the *one*-dimensional eigenstates expressed as functions of one or another of the coordinates x, y, and z. Let E_n be the corresponding one-dimensional energies. Then the eigenfunctions of the three-dimensional oscillator problem [call them $v(x, y, z)$], are labeled by three integers, n_1, n_2, and n_3. The eigenfunctions and corresponding energies are

$$\begin{aligned} v_{n_1,n_2,n_3} &= u_{n_1}(x)u_{n_2}(y)u_{n_3}(z); \\ E_{n_1,n_2,n_3} &= E_{n_1} + E_{n_2} + E_{n_3} = \hbar\omega\left(n_1 + n_2 + n_3 + \frac{3}{2}\right), \end{aligned} \tag{5.10}$$

where the integers again range from zero to infinity. Notice that the energy depends on these integers only through their sum, which itself is, of course, an integer. We may therefore label the energies with a single index n defined by $n = n_1 + n_2 + n_3$:

$$E_n = \hbar\omega\left(n + \frac{3}{2}\right), \quad n = 0, 1, 2, 3, \ldots \tag{5.11}$$

The situation is one of degeneracy, since except for $n = 0$ there are different ways to partition the integer n into a sum of three non-negative integers n_1, n_2, n_3. For the ground level, $n = 0$, there is no degeneracy; uniquely $(n_1, n_2, n_3) = (0, 0, 0)$. But for

$n = 1$ there are three partitions: $(1, 0, 0)$, $(0, 1, 0)$ and $(0, 0, 1)$. For $n = 2$ there are six partitions (please check!), and so on, with the degeneracy increasing as the energy index n gets larger and larger. Each different partition for a given n corresponds to a different eigenfunction.

The ground level of the spherical oscillator has energy $3\hbar\omega/2$. Its wave function, as we see from Eqs. (5.8) and (5.10), is

$$v_{0,0,0} = N^3 \exp\left(-\frac{r^2}{2r_0^2}\right), \qquad r_0 = x_0 = \sqrt{\frac{\hbar}{m\omega}}. \tag{5.12}$$

In this three-dimensional context we have renamed x_0, calling it r_0.

Central Potentials Generally

A potential $V(r)$ is said to be central if it depends on x, y, and z only in the combination r, where r is the distance to the origin. The potential may be said to be "centered" at the origin. The force described by a spherical potential is a *central force*; it points in the radial direction with strength $F = -dV/dr$. A positive value of F indicates that the force is repulsive, that it points outward; negative F corresponds to an attractive force, one that points to the origin. Of course, it can happen that the force is repulsive for some ranges of r, attractive for others. The spherical oscillator potential discussed above is an example of a purely attractive central potential.

A central potential has no preferred direction in space. The physics associated with it possesses rotational symmetry; it is invariant under arbitrary rotations about an arbitrary axis passing through the origin. This symmetry property has important implications. Classically, it entails conservation of angular momentum **L**: the angular momentum of a particle moving in a central potential remains unchanged in magnitude and direction as it traverses its orbit. This in turn implies that the classical orbit must lie in a plane, with **L** normal to the plane. All orientations of the plane of motion are possible. The orientation for any particular orbit is determined by initial conditions.

Rotational symmetry also implies that all orientations of an orbit *within* a plane are equally allowed, the particular orientation depending on initial conditions. For example, the earth moves around the sun in a particular elliptical orbit (which happens to be nearly, but not quite, circular). The major axis of that ellipse points in a particular direction in space. The central force of gravity would have permitted it to point in any other direction in the plane; and indeed would have permitted any other orientation of the plane.

The classical situation may be put more generally in the following way. Given some potential V, whether central or not, the Newtonian laws of motion encompass infinitely many different orbits. Which orbit a particle occupies is determined by initial conditions. What is entailed by some geometric symmetry, if there is one, is a relation among the orbits. In the case of rotation symmetry, if you know any one orbit, you know others obtained from it by arbitrary rotations, as described above. That's a powerful insight.

The quantum mechanical equivalent of classical angular momentum conservation is the statement that all three Cartesian components of the angular momentum observable **L** commute with the energy observable. As already discussed, the three components do not commute among themselves, but L^2 does commute with the component of **L** in any direction. For a particle moving in a central potential, we can therefore find states that are simultaneous eigenstates of energy as well as of L^2 and the component of **L** in any direction of our choice. Let us take that direction to be the z axis. The simultaneous eigenstates will now bear the quantum numbers ℓ and m_ℓ [see Eqs. (4.21) and (4.22)]. For given values of these quantum numbers there will be a spectrum of energies. For notational simplicity, suppose that the spectrum is discrete. We can then introduce a principal quantum number n (really, just a counting index) to distinguish among linearly independent states having the same quantum numbers ℓ and m_ℓ. The simultaneous eigenstates may thus be written u_{n,ℓ,m_ℓ}. The energy corresponding to this state we provisionally denote by E_{n,ℓ,m_ℓ}.

In fact, one can easily show for a central potential that the energy does *not* depend on m_ℓ; namely, that there is degeneracy in this quantum number. The energies $E_{n,\ell}$ thus depend only on the two indices n and ℓ. The $2\ell+1$ states u_{n,ℓ,m_ℓ} having the same indices n and ℓ but differing in m_ℓ all have the same energy. This degeneracy is the quantum analog of the classical result that the angular momentum **L** can point in any orientation. The quantum degeneracy in m_ℓ follows from the fact that the central potential has no favored direction in space.

Let us be clear about the principal quantum number n. Consider all the linearly independent states having a given pair of quantum numbers ℓ and m_ℓ in common. The states in this set will typically all have different energies. Now attach a counting number n to distinguish among the states, such that n increases with increasing energy. The decision about where to start the counting from—what number n_{min} to assign to the lowest energy—is a matter of convenience and convention. It is sometimes best to make different choices of n_{min} for different ℓ values.

The eigenstates are best expressed in spherical coordinates. In these coordinates the eigenfunctions have the structure

$$u_{n,\ell,m_\ell} = R_{n,\ell}(r) Y_\ell^{m_\ell}(\theta, \phi), \tag{5.13}$$

where the spherical harmonic factors, the functions of polar angle θ and azimuth ϕ, guarantee that the solution is an eigenstate of L^2 and L_z. When this is inserted into the energy eigenvalue equation (4.1), one finds an ordinary differential equation for the radial function R; or better, for the product rR:

$$\frac{d^2(rR)}{dr^2} + \frac{2mE}{\hbar^2}(rR) = \frac{2mV(r)}{\hbar^2}(rR) + \frac{\ell(\ell+1)}{r^2}(rR). \tag{5.14}$$

We have temporarily suppressed the indices n and ℓ on the radial function R. The well-behaved solutions of Eq. (5.14) depend on the angular momentum quantum number ℓ (but not on m_ℓ, which does not appear in the above equation) and are distinguished one from another by the principal quantum number n; hence $R \to R_{n,\ell}$. Similarly $E \to E_{n,\ell}$.

As usual, you will not be asked here to solve this equation in the forward direction for any particular potential V. Indeed, for most cases of real interest, no simple, analytic solutions are available and one has to resort to numerical or approximation methods. But where an analytic solution is presented below, you may just possibly want to try to confirm it!

The so-called *radial equation* (5.14) is similar to the energy eigenvalue equation for a particle moving one-dimensionally in a potential $V(x)$, with the following differences: (1) The variable x is replaced by the variable r, which of course ranges only over non-negative values; and the one-dimensional eigenfunction $u(x)$ is replaced by the product function rR. This product must vanish at the origin since r vanishes there. Hence, in one-dimensional language, it is as if there is a wall at $x = 0$, with x ranging only over non-negative values. (2) In addition, in one-dimensional language it is as if the potential $V(x)$ is replaced by $V(x) + \ell(\ell+1)\hbar^2/2mx^2$. The extra term represents a centrifugal force effect.

The One-Electron Atom

This has been the great spawning ground of quantum mechanics, from Bohr to Schroedinger to Dirac to the Lamb shift and quantum electrodynamics. By the one-electron atom we mean the "hydrogenic" atom, a system composed of one electron and one nucleus: the true hydrogen atom, once-ionized helium, twice-ionized lithium, and so on. For Schroedinger's opening salvo in quantum mechanics, as in Bohr's "old" quantum theory, it was good enough to ignore various refinements, treating the electron as a nonrelativistic particle subject only to the Coulomb attraction of a point nucleus. This brings one to close, but by no means perfect, agreement with experiment. For example, in the ground level of hydrogen the ratio of the root mean square electron velocity to the speed of light is equal to $1/137$. That's a small enough number to justify the expectation that relativistic corrections will be small, as indeed they are; but they are not negligibly small. Then, there is the fact that the electron has spin. This in itself would not change the energy levels if there were no spin-dependent forces. But there *are* such forces, and these produce shifts on about the same scale as do the relativistic corrections. Rather than deal with these corrections piecemeal and imperfectly, Dirac set out to produce a fully relativistic equation for the electron. He was guided by considerations that left electron spin and the nature of spin forces as open questions. The answers emerged from his equation on their own, and with spectacular success. But the triumph was not quite total. There are tiny remaining discrepancies with experiment, although these were not definitively detected until almost two decades later. Their resolution involved and gave credence to the principles of quantum electrodynamics, the relativistic quantum field theory of electrons and photons. More on that later.

For now let us return to the humble, nonrelativistic atom, which consists of a single electron of mass m and charge $-e$ moving around a fixed point nucleus of charge Ze. For the moment, neglect spin. The Coulomb potential is $V(r) = -Ze^2/r$. It falls to zero as r grows large, so we know that the energy spectrum is continuous for positive energies, $E > 0$. Here we will be concerned only with the bound states, $E < 0$. Since the potential is central we can invoke Eqs. (5.13) and (5.14) for the energy eigenvalue problem. The radial equation, to say it once again, has solutions for *any* value of the energy E; but typically the solutions are ill behaved. However, at certain energies E, the eigenvalues, there is one acceptable solution. For each value of the angular momentum quantum number ℓ there is a spectrum of energies and corresponding radial functions. We provisionally introduce a counting index denoted by upper-case N, where N ranges upward from $N_{\text{min}} = 0$ whatever the value of ℓ. The bound state energy spectrum for given quantum number ℓ is then found to be

$$E_{N,\ell} = -\frac{Z^2 e^4 m}{2\hbar^2} \frac{1}{(N+1+\ell)^2}, \qquad N = 0, 1, 2, 3, \ldots$$

Notice that the energy depends on the integers N and ℓ only in their sum. We can therefore usefully define an integer quantum number $n = N + 1 + \ell$. For given ℓ, n ranges upward from $n_{\text{min}} = \ell + 1$. Turning this around, for given n, the quantum number ℓ ranges from 0 on up to $\ell_{\text{max}} = n - 1$. In this latter way of expressing the matter, the bound state energies are

$$E_n = -\frac{Z^2 e^4 m}{2\hbar^2} \frac{1}{n^2}, \tag{5.15}$$

where $n = 1, 2, \ldots, \infty$; and, for given n, $\ell = 0, 1, 2, \ldots, n-1$.

The corresponding eigenstates u_{n,ℓ,m_ℓ} are labeled with three indices. Notice the degeneracy situation. The energy E_n does not depend on the quantum number m_ℓ. That is as expected; it is the case for any central potential. What is not true for the random central potential, however, is that there is also degeneracy in the quantum number ℓ. This is peculiar to the Coulomb and spherical oscillator potentials. For a given energy E_n, ℓ can take on any of the values indicated above; for each ℓ, m_ℓ ranges in unit steps from $-\ell$ to ℓ. In the ground level ($n = 1$), ℓ has uniquely the value $\ell = 0$, so there is no degeneracy here. For $n = 2$ there are two ℓ values: $\ell = 0, 1$. For $\ell = 0$, m_ℓ is uniquely equal to zero. But for $\ell = 1$, $m_\ell = -1, 0, 1$. Altogether then, the $n = 2$ energy level is fourfold degenerate. It is easy to work out the general case. The degeneracy d_n of the nth level is $d_n = n^2$. This result follows from summing up the quantity $2\ell + 1$ over all values of ℓ from 0 to $n - 1$. But looking ahead, keep in mind that all of this so far is without electron spin.

We will write out here only the ground state wave function. It is very simple:

$$u_{\text{gnd}} \equiv u_{1,0,0} = \frac{1}{\sqrt{\pi a^3}} \exp(-r/a), \quad a = \frac{a_B}{Z}, \quad a_B = \frac{\hbar^2}{me^2}. \tag{5.16}$$

Up to a multiplicative constant the above solution is just the radial function $R_{1,0}$. You are invited to confirm that it in fact solves the radial equation with E set equal to the ground state energy. The solution falls off exponentially with increasing r, being mainly concentrated in a volume of radius equal to the parameter a. That parameter is the Bohr radius divided by the nuclear charge parameter Z. To characterize the size of the atom when it is in a state labeled by the quantum numbers n and ℓ, it is convenient to adopt the reciprocal of the expectation value $\langle 1/r \rangle$ as a measure of the inverse size. With this, it happens to be the case that the size parameter depends only on n. For the nth energy level, the result is

$$\langle 1/r \rangle_n = 1/n^2 a. \tag{5.17}$$

Based on this measure, the size of the atom in the nth energy level may be taken to be $(n^2/Z)a_B$. Recall that the Bohr radius is $a_B = 0.53 \times 10^{-8}$ cm.

It may be instructive to digress here in order to pursue some *dimensional* reasoning. The energy eigenvalue problem for the one-electron atom involves only two input parameters: Ze^2 and the ratio $\hbar^2/m$. The atomic charge Z is dimensionless, that is, a pure number (Z = 1 for hydrogen, Z = 2 for helium, etc.). Since e^2/r is an energy, e^2 has the dimensions [energy]·[length]. Planck's constant $\hbar$ has the dimensions [energy]·[time]. Mass has the dimensions [energy]·[time]^2/[length]^2. Putting these things together, we can check that our two parameters have the following dimensions:

$$[Ze^2] = [\text{energy}] \cdot [\text{length}]; \quad [\hbar^2/m] = [\text{energy}] \cdot [\text{length}]^2.$$

Thus, the only energy parameter in this problem is $(Ze^2)^2/(\hbar^2/m) = Z^2e^4m/\hbar^2$. Energy levels have no choice but to be equal to dimensionless numbers multiplied by this quantity, as is borne out by Eq. (5.15). Similarly, any quantity with the dimension of length has to be a dimensionless number multiplied by the ratio $\hbar^2/me^2$. The formula for the Bohr radius bears this out as well. All of this can be anticipated in advance, so that the eigenvalue problem reduces to finding those dimensionless numbers. The reader is invited to go through similar dimensional reasoning for the harmonic oscillator problem.

Let us now turn to some refinements to the hydrogenic atomic. One of them is easily dealt with. Until now we have been treating matters as if we were dealing with a one-body problem. The nucleus has been regarded as fixed, as if infinitely massive. Its only role was to provide the Coulomb field in which the electron moves. Fortunately, in quantum just as in classical mechanics, it is easy to take the finiteness of the nuclear mass into account and deal properly with the two-body nature of the problem. This is so if, as here, the force between the two bodies depends only on their relative separation. We have merely to recognize that everything we have been doing refers not to a fixed laboratory frame but to the center of mass frame of electron and nucleus, and the energy levels are referred to that frame. The two-body problem reduces effectively to the one-body problem we have been considering with this one change: the mass m in all our formulas is no longer the electron mass, it is the *reduced mass* of the two-body system,

$$m = \frac{m_e M_n}{m_e + M_n} = m_e \cdot \frac{1}{1 + m_e/M_n},$$

where m_e is the electron mass and M_n the nuclear mass. Since the former is much smaller than the latter, the reduced mass is not very different from the electron mass. Even for hydrogen, the difference is only one part in a little less than two thousand. Nevertheless, spectroscopists are quite able to detect such refinements. For example, if we could ignore the reduced mass effect, it would follow from Eq. (5.15) that the $n = 1$ level of hydrogen ($Z = 1$) would have an energy that exactly coincides with that of the $n = 2$ level of once-ionized helium ($Z = 2$). However, discrepancies were inferred experimentally early on after Bohr's model was introduced; and it was Bohr himself who recognized that these differences arise from the difference in reduced masses of the hydrogen and once-ionized helium atoms.

A deeper refinement to our treatment of the one-electron atom arises from relativistic effects. Recall that the correct formula for the kinetic energy of a particle of mass m and momentum p is, according to Eq. (2.14),

$$K = E - mc^2 = \sqrt{(mc^2)^2 + (cp)^2} - mc^2.$$

For particle velocities v small compared to the speed of light c, $v/c = cp/mc^2 \ll 1$. To leading order in this small ratio, the kinetic energy then reduces to the familiar expression $K = p^2/2m$. In next approximation, one picks up a correction term $-p^4/8m^3c^2$. This can be inserted as an added operator term in the energy eigenvalue equation; and it is not hard to work out to lowest order the small energy shifts that it induces. These corrections were computed early on after the birth of the new quantum mechanics, although they had in fact already been figured out in the framework of the old quantum mechanics. Either way, the relativistic treatment was not full and exact. Rather, as described just above, the relativistic correction was treated in first order as a small perturbation. The corrections are indeed small.

Next, there is electron spin to be considered. The very fact of spin increases the space of quantum mechanical states. The most general spin state is a linear combination of states with spin "up" along some chosen z axis and "down" along that axis; in the notation of Eq. (4.24), $m_s = +1/2$ and $-1/2$, respectively. For greater brevity, we may denote these two spin states with the symbols $\uparrow$ and $\downarrow$. Suppose the electron spin is purely up at all points in space. Then, for its wave function we write $\Psi = f(\mathbf{r}, t)\uparrow$, where the space-time function f is normalized to unity and where f^*f has the usual interpretation as the spatial probability density for the (spin-up) electron. If the spin is purely down at all points in space, we write $\Psi = g(\mathbf{r}, t)\downarrow$. The actual wave function will in general be some linear combination

$$\Psi = af(\mathbf{r}, t)\uparrow + bg(\mathbf{r}, t)\downarrow,$$

where a and b are constants, normalized to $a^*a + b^*b = 1$. The spatial probability densities for spin up and spin down are, respectively, a^*af^*f and b^*bg^*g. The relative probability, independent of spatial position, is simply a^*a/b^*b.

Now let us turn to the energy eigenvalue problem and suppose to start with that the force acting on the electron is spin independent; namely, that although the electron has spin, the force does not care about the spin. In this situation the spin observable $\mathbf{S}$ obviously commutes with the energy observable and we can therefore find simultaneous eigenstates of energy and the component of $\mathbf{S}$ along any axis, say the z axis. It is clear that the energy eigenvalues obtained without any consideration of spin will be unchanged when we take spin into account. However, the number of eigenstates will double. Namely, suppose that in the absence of spin considerations we had found an eigenstate $u(x, y, z)$ corresponding to energy E. With spin taken into account, both $u\uparrow$ and $u\downarrow$ will now be eigenstates with that *same* energy E. Suppose that the particle is moving in a central potential. In the absence of spin considerations, it will have simultaneous eigenstates of energy, L^2 and L_z. We have denoted these by u_{n,ℓ,m_ℓ}. Let $E_{n,\ell}$ be the corresponding energies (for hydrogen there is degeneracy in ℓ; but that's atypical for central potentials). With spin allowed for, the states acquire an extra index m_s, so that we now write u_{n,ℓ,m_ℓ,m_s}. For example, for spin up, $u_{n,\ell,m_\ell,1/2} = u_{n,\ell,m_\ell}\uparrow$; and relatedly for $m_s = -1/2$. The energy, however, is independent of the spin quantum number m_s.

The Coulomb force is spin independent and therefore does not have an effect on energy levels. How is it then that spin shows up in spectroscopy, as it indeed does? The answer is that there *do* exist forces that are spin dependent and that can therefore remove the spin degeneracy. To set the stage, first take the case of an assumed *spinless* electron moving in some central potential; and consider what happens when one subjects it also to a uniform magnetic field $\mathbf{B}$. In the energy eigenvalue equation, the influence of the magnetic field shows up as a term proportional to the product of the field strength B and the component of angular momentum in the direction of the field. For convenience, take the z axis to lie along the direction of the field. In the eigenvalue equation (4.1) this new term, which is to be added to the potential V, is

$$\frac{eB}{2mc}L_z \tag{5.18}$$

From Eq. (4.22) we see that the "perturbed" eigenvalue equation (the equation in the presence of the B field) has the same eigenfunctions u_{n,ℓ,m_ℓ} as the "unperturbed" equation. The field does not change the eigenfunctions. But the energies are shifted by the amount $eB\hbar m_\ell/2mc$, thus lifting the degeneracy in the quantum number m_ℓ. That is, the energy levels E'_{n,ℓ,m_ℓ} in the presence of the field (we mark them with a prime) depend on m_ℓ and are related to the unperturbed energies $E_{n,\ell}$ by

$$E'_{n,\ell,m_\ell} = E_{n,\ell} + \frac{e\hbar B}{2mc} m_\ell.$$

The energy level that was degenerate in m_ℓ in the absence of the magnetic field now separates into a group of $(2\ell + 1)$ sublevels of different energies. Because a magnetic field has this effect, the angular momentum quantum number m_ℓ is often called the (orbital) *magnetic* quantum number. The shift of atomic energy levels in a magnetic field is known as the *Zeeman effect*.

All of the above was already anticipated in the old quantum theory, in a spinless era. The discovery of spin, which occurred around the time of the birth of the new quantum theory, was triggered in part by problems that had been encountered with the Zeeman effect described above. The spin hypothesis was devised to resolve these problems. Spin effects come about in the following way. Equation (5.18) describes an energy term that arises from the interaction between magnetic field and orbital angular momentum. If you assume that the electron has spin, it seems natural to conjecture that there is a similar interaction between the magnetic field and the spin angular momentum, a term exactly like that in Eq. (5.18) but with L_z replaced by S_z. To play it safe—since this is only a conjecture at the outset—let us multiply by a phenomenological factor g_e to be determined by experiment. The added energy term representing the interaction of the spin with the B field is thus

$$g_e \frac{eB}{2mc} S_z.$$

The parameter g_e is the so-called *Landé g factor*, the subscript indicating that we are dealing here with an electron.

Altogether then, with both kinds of interactions included, the perturbed energies are related to the unperturbed energies by

$$E'_{n,\ell,m_\ell,m_s} = E_{n,\ell} + \frac{e\hbar B}{2mc}(m_\ell + g_e m_s). \tag{5.19}$$

In the absence of the magnetic field, a level of given quantum numbers n and ℓ is $2(2\ell + 1) =$ degenerate, the first factor 2 representing the number of different m_s values and the second factor the number of m_ℓ values. The effect of the magnetic field is to split this unperturbed level into a group of sublevels, with energies labeled as above by the quantum numbers m_ℓ and m_s.

Early on it became known experimentally that the value of the g factor is $g_e = 2$, to within the measurement uncertainties. Before the advent of Dirac's relativistic equation for the electron, that value had simply to be accepted as an empirical fact. One of the several great triumphs of the Dirac equation was that precisely this value emerged automatically. The integer 2 precisely! That triumph is undiminished by the fact that this is not quite exactly the empirical value. We now know from experiments of astonishing precision that

$$g_e = 2 \times (1.001159652193 \pm 0.000000000010). \tag{5.20}$$

The very tiny departure from Dirac's whole number arises from quantum field-theoretic effects. These can be calculated theoretically to equally astonishing precision and turn out to be in full agreement with experiment! For now, however, let us accept the whole number 2 as a very good practical approximation, and return to the pre-Dirac atom but with small relativistic and spin effects taken into account in first order.

There is one further spin-dependent effect to be dealt with, this one having nothing to do with an external magnetic field. It arises as follows. In the reference frame of the nucleus, and in the absence of any *external* magnetic field, the only electromagnetic field that the electron experiences is the Coulomb field of the nucleus. But let us now imagine ourselves sitting on the electron. In accordance with the relativistic transformation equations discussed in Chapter 2, in that moving electron frame there is not only a very slightly modified Coulomb electric field, but also a nonvanishing *magnetic* field. We may then expect an interaction term between that magnetic field and the electron's spin, just of the sort discussed above when the field was an external one. The net effect of all of this is that there is the following additional term to be included in the Hamiltonian:

$$\xi(r)\{L_x S_x + L_y S_y + L_z S_z\},$$

where the function ξ depends on the choice of central potential V. For the hydrogenic case

$$\xi(r) = \frac{Ze^2}{2m^2c^2r^3}.$$

Since the above energy term involves both orbital and spin angular momentum observables, it is called the *spin-orbit interaction*. It is amusing that the relativistic reasoning that goes into determining $\xi(r)$ has some delicacies that a rash derivation can overlook. Einstein himself got the numerical factor wrong initially. L. J. Thomas got it right.

With the spin-orbit interaction term included in the energy eigenvalue equation, it is no longer true that energy commutes with L_z and S_z. But it *does* commute with L^2 and with the total angular momentum observable $\mathbf{J} = \mathbf{L} + \mathbf{S}$; therefore with J^2 and any component of $\mathbf{J}$, say the J_z component. To remind yourself about total angular momentum, see Eqs. (4.26) and (4.27). Altogether then, one can organize the eigenstates of energy to be, simultaneously, eigenstates of L^2, J^2, and J_z. The energy eigenstates (call them v_{n,ℓ,j,m_j}) are thus labeled by the orbital angular quantum number ℓ, the total angular momentum quantum numbers j and m_j, and a principal quantum number n. From Eq. (4.28) we know that, for any given j, the orbital quantum number ℓ can take on only two values: $\ell = j + 1/2$, $j - 1/2$. Because there is no preferred direction in space, we can be sure for given n, ℓ, j that the energy will not depend on the quantum number m_j. The energies $E_{n,\ell,j}$ are therefore labeled only by the three quantum numbers indicated and the degree of degeneracy is $2j+1$. A set of degenerate $2j+1$ states sharing the same quantum numbers n, ℓ, j but differing in m_j constitutes what spectroscopists call a *multiplet*.

All of this is for an arbitrary central potential. But the Coulomb potential is special. It has additional degeneracy. We already encountered this in the zeroth-order treatment of the one-electron atom, before relativistic or spin-orbit corrections made an appearance. In that zeroth-order approximation there is degeneracy in ℓ for given n: the energy levels depend only on the principal quantum number n. Something of this persists with the above corrections taken into account. For given n, the energies now vary from one j value to another so that there is a splitting into sublevels, with j ranging in unit steps from $j = 1/2$ to $j = n - 1/2$. But for given j there is degeneracy in ℓ: both $\ell = j + 1/2$ and $\ell = j - 1/2$ have the same energy. In summary, for the one-electron atom the energies $E_{n,j}$ depend only on the two quantum numbers indicated. The degeneracy is $2(2j+1)$, where the second factor represents the degeneracy associated with the quantum number m_j and the factor 2 in front comes from the number of possible ℓ values for given j. Here then are the energy eigenvalues:

$$E_{n,j} = -\frac{Z^2 e^4 m}{2\hbar^2}\left\{1 + \frac{(Z\alpha)^2}{n^2}\left(\frac{2n}{2j+1} - \frac{3}{4}\right)\right\},$$
$$\alpha = \frac{e^2}{\hbar c} \approx \frac{1}{137}. \tag{5.21}$$

The expression in front of the curly brackets is the zeroth-order result. The expression in the curly brackets is equal to unity plus a correction term. The correction is of order $(Z\alpha)^2$, which is small enough if the atomic number Z is not too large. It happens that Dirac's equation for the relativistic electron can be solved exactly for the hydrogenic atom. The exact result preserves the main qualitative feature obtained above, namely that there is degeneracy in ℓ for given j. Moreover, to first order in $(Z\alpha)^2$ the Dirac result agrees with the above formula, but it goes on to include all higher-order corrections as well. For small Z those higher-order corrections are very small.

Absolute energies are harder to measure experimentally than energy differences. Of special interest, then, is the question whether there is any energy separation between states with the same n and j quantum numbers but different ℓ values. In particular, consider the $n = 2$ states of hydrogen ($Z = 1$). There are three multiplets here: $(j, \ell) = (1/2, 0)$, $(1/2, 1)$, and $(3/2, 1)$. Their respective spectroscopic designations are $s_{1/2}$, $p_{1/2}$ and $p_{3/2}$. The subscript tells the value of j; and for reasons we need not go into here, the letter s stands for $\ell = 0$, p for $\ell = 1$. Dirac's prediction is that the $s_{1/2}$ and $p_{1/2}$ multiplets have *exactly* the same energy; moreover, that this pair should lie below the $p_{3/2}$ level, with the separation predicted by Eq. (5.21)—or even more precisely by the fully relativistic Dirac formula.

The more sensitive of these two tests concerns the predicted degeneracy of the $s_{1/2}$ and $p_{1/2}$ multiplets. What does experiment say? It was not until the middle 1940s, long after Eq. (5.21) had been promulgated and then given a firmer foundation by Dirac, that the first sure detection of a separation between these multiplets occurred. This was the work of W. Lamb and R. Retherford. What they found is that the $p_{1/2}$ level lies below the $s_{1/2}$ level by somewhat less than 10^{-5} electron volts. It is common in spectroscopy to express energy differences ΔE in terms of the frequency f of an imaginary photon carrying that energy: $f = \Delta E/2\pi\hbar$, where ΔE is the energy difference and f is the ordinary repetition frequency. The *Lamb shift*, as it is called, is now known experimentally to a high level of accuracy:

$$\text{Lamb shift} = 1057.86 \text{ megacycles/sec.}$$

As with the anomalous Landé g factor of Eq. (5.20)—"anomalous" in that it departs from the prediction of the Dirac equation—the existence of the Lamb shift has its origins in the quantum field theory of electrons and photons. As soon as Lamb reported the original findings, the field theorists set to work and were able to give a good account of the matter. The experimental and theoretical accuracies have increased substantially in subsequent years and agreement continues to survive.

We have devoted a lot of space to the one-electron atom because it has played such a central role in the development of quantum mechanics. A lot more could be said: for example, about the level shifts induced by electric fields (Stark effect), about the Zeeman effect that we only touched on earlier, and so on. The Zeeman effect formula of Eq. (5.19) is in good agreement with experiment for strong magnetic fields, but for weak fields it runs into trouble. The picture is confounded by the influence of spin-orbit coupling, which Eq. (5.19) neglects. In the weak field region one speaks of the "anomalous" Zeeman effect. It was a great puzzle in the early days of quantum mechanics, before the advent of spin-orbit coupling; but soon enough everything fell into place.

There is one last hydrogenic topic to be touched upon here. We have been treating the atomic nucleus so far as a geometric point. Actually, the neutrons and protons that make up a nucleus are spread out (in a quantum probabilistic sense) over some volume characteristic of the nucleus. Very roughly, the radius is $R \approx A^{1/3} \times 10^{-13}$ cm, where A is the total number of neutrons and protons. The size of the one-electron atom is roughly a_B/Z, where $a_B = 0.53 \times 10^{-8}$ cm. Even for large nuclei, the chance of finding the electron inside the nucleus is tiny, so it is a good approximation to treat the nucleus as a point particle, as we have done. But now consider the negative muon. It is a particle with the same charge, spin, and many other properties as the electron—except for two things: (1) it is unstable, and (2) it is about 200 times heavier than the electron. When a muon traveling through a medium is captured into an atomic orbit its Bohr radius is about 200 times smaller than that of the electron. So in a muonic atom, especially if the nucleus has large atomic number A and large atomic charge Z, the muon spends a lot of time inside the nucleus. There the potential V is no longer that produced by a single point charge. In fact, the potential energy there resembles that of a spherical oscillator. The two exemplary potentials so beloved of pedagogs therefore come into play: the oscillator for $r < R$ and the Coulomb potential for $r > R$, where R is the nuclear radius.

The Infinite Solenoid

Within a few short years after the foundation of quantum mechanics, its basic principles and peculiarities had become well absorbed. Yet even in its simplest precinct, the nonrelativistic motion of a single particle, there have been surprises. In particular, we shall discuss here a strange effect first noticed and expounded in the work of Y. Aharonov and D. Bohm more than three decades later.

Consider a solenoid. A solenoid is a long circular cylinder with turns of a current-carrying wire wrapped helically around it along its whole length. An ideal solenoid is one that is infinitely long. For an infinite solenoid the current generates a magnetic field that is confined solely to the interior of the cylinder. The field inside points along the solenoid and is uniform in strength throughout the interior. The important thing about the ideal solenoid is that there is *no magnetic field on the outside.* Now consider such a solenoid surrounded on the outside by a concentric, cylindrical wall. Its role, as ideal wall, is to keep any particle placed on the outside of the wall from penetrating into the solenoid. In short, a charged particle placed outside the wall, governed there by the rules of quantum mechanics, has zero probability of being found inside the solenoid; and thus zero probability of directly experiencing the magnetic field that is confined to the inside.

But simple quantum mechanical calculations show, and experiments confirm, that the quantum behavior of a charged particle on the outside in fact responds to changes in the magnetic field strength on the inside! Let us illustrate this with an example that is simple enough to permit explicit, analytic calculations. For the purposes of this example we can remove the concentric cylindrical wall described above and replace it by a hollow torus (doughnut) that is concentric with, and lies outside of, the solenoid. Take the wall material of the torus to be ideal, so that a particle placed inside the doughnut cannot penetrate to its outside, not even quantum mechanically. Thus, as before, the particle surely cannot penetrate to the inside of the solenoid. Suppose the particle is charged. Our intuition must then be that the particle, though it would certainly be influenced by a magnetic field if it experienced one, cannot know that there happens to be a field inside the solenoid. To check this intuition, let us consider the energy levels of the particle moving inside the torus. Idealizing further, take the torus to be a circular loop of very thin, hollow material (a hollow noodle, say). In the limit of extreme thinness, the formula for the energy levels takes on a very simple form. Let Q be the particle's charge, M its mass, and R the radius of the loop of pasta encircling the solenoid. One then finds that the energy levels are given by

$$E_m = \frac{\hbar^2}{2MR^2}\left(n - \frac{QF}{2\pi\hbar c}\right)^2, \quad n = 0, \pm 1, \pm 2, \pm 3, \ldots \qquad (5.22)$$

Here F is the magnetic flux through the solenoid, the product of magnetic field strength B and cross-sectional area of the solenoid.

Unmistakably, the energies depend on the magnetic flux F and hence on the magnetic field; yet the particle is confined quantum mechanically to a field-free region! Of course, the problem has been idealized. One essential assumption is that the solenoid is ideal. Another is that the wall of the torus looping around the solenoid is impenetrable. The added assumption of an infinitely thin torus is not essential; it was made merely to simplify the energy level formula. Idealizations (gedanken experiments) are a legitimate and honorable tradition in quantum mechanics. Moreover, one can in practice construct solenoids that are close enough to ideal, with very little leakage of magnetic field to the outside; and one can create walls that are close enough to ideal. There is one more interesting thing to be noticed here. The energy levels shift with changing magnetic flux F, but notice that the pattern repeats itself if F is replaced by

$F + (2\pi\hbar c/Q)N$, where N is any integer. The quantity $2\pi\hbar c/Q$ is the so-called magnetic flux quantum.

So what's going on? The answer is that quantum mechanics is strange. This solenoidal oddity is special to magnetic fields. The phenomena described here would not occur if the magnetic field were replaced by an electric field confined to the inside of the solenoidal cylinder. In that case, the charged particle on the outside would be indifferent to the presence of the field on the inside. Somehow the magnetic field conveys information to the space beyond its direct reach. The relevant property of that space has to do with its topology. Consider the space that lies outside the infinite, cylindrical solenoid. In that world you can imagine forming loops of string that can be drawn as tight as you please, shrunk down to a point, without penetrating the solenoid. But there are other loops, those encircling the cylinder, that cannot be shrunk indefinitely in this way. On that account the space outside the cylinder is said to be "multiply connected."

Now, this may be very interesting to a topologist, but does the magnetic field care? The answer is yes; in the quantum mechanical context the magnetic field cares. Alas, it is not easy to go beyond this elliptical statement without becoming unduly technical.

Decay Processes

The term "radioactivity" was first coined in connection with the α, β, and γ nuclear decay reactions discussed in Chapter 1. In an α decay reaction, the nucleus transforms spontaneously into a different (daughter) nucleus containing two fewer protons and two fewer neutrons, these having become bound together as an α particle (helium nucleus) and then ejected. A fully detailed quantum analysis is quite complicated, but at least there is no need to appeal to particle creation or destruction for this particular kind of radioactivity. The ingredients of the alpha particle preexist in the parent. What happens in the decay process is that they somehow assemble and then get kicked out. In contrast, for β decay the ejected electron and neutrino do not preexist in the parent nucleus; instead, they are spontaneously created as a neutron in the nucleus decides to undergo the decay $n \to p + e + \nu$. In the process, the nucleus transforms to a daughter with one less neutron and one more proton. So too in γ decay, it is not profitable to think of the photon as preexisting in the nucleus. Rather, one deals here with spontaneous creation, the photon being created as the nucleus decides to jump from an excited quantum level to a lower level. The same holds for photon emission when an orbital electron in an atom jumps from an excited to a lower level. That is, nuclear and atomic *radiative* transitions (photon emission) are of the same ilk, though the photon energies are on different scales in the two cases, typically much larger in the nuclear case. In these photon processes, there is no change of nuclear or atomic species, but the energy levels (of the nucleus in the one case, the orbital electron system in the other) have changed. Finally, at the subnuclear level—the world of various kinds of mesons, baryons, leptons, gauge bosons—most particle species are unstable, each with its own characteristic decay modes and mean lifetimes. The underlying dynamics for much of this is a topic at the frontiers of contemporary particle physics.

In this whole array of decay processes, with the exception of α particle decay, the appropriate language is that of particle creation and destruction; and the appropriate theoretical framework that of quantum field theory, which we have not yet touched. The process of α particle emission stands nearly alone in that it lends itself to treatment in the framework of particle quantum mechanics. It is explicable in terms of tunneling. Before we take that up, however, let us make some observations of a very general character about decay processes, whether atomic, nuclear, or subnuclear.

Shortly after the discovery of α, β, and γ radioactivity, and well before the Rutherford model of the atom was proposed, Rutherford and Soddy introduced a kind of actuarial analysis that has prevailed ever since. Take a sample of some radioactive species and let $N(t)$ be the number of parent atoms still surviving at time t. Let ΔN be the net change in N in the time interval between t and $t + \Delta t$, where Δt is a tiny, positive time increment. Clearly ΔN will be negative; and it seemed plausible to Rutherford and Soddy that it should be proportional to Δt and also to the number of still-surviving parents in the sample, $N(t)$. Passing to the limit of differentials, the hypothesis is that

$$dN(t) = -N(t)dt/\tau,$$

where the proportionality constant $1/\tau$ is a parameter characteristic of the parent species. This equation is easily solved. Let $N(0)$ be the number of parent atoms present at the initial time $t = 0$. Then at later time t the number of survivors is

$$N(t) = N(0)\exp(-t/\tau). \quad (5.23)$$

This is the promised exponential decay law. It is easy to check that the average (or mean) lifetime is τ.

Several qualifications should be noted here. We have assumed that the population changes with time only because parents decay. If the parents are themselves the daughters of grandparent species, the parental population will grow from the one cause, decrease from the other. The analysis is not hard to carry out though we will forgo it here. Another qualification is that we have treated $N(t)$ as if it is a continuous variable, although realistically it is always a whole integer, decreasing by one whole unit every time a parent decays. This is not a serious error, however, as long as $N(t)$ is very large compared to unity. If the formula says that there are 1,000,000,000.7 parents left at a certain instant, feel free to round to a nearby whole number.

One of the earliest applications of quantum ideas to the nucleus was made in connection with the phenomenon of α particle radioactivity. The protons and neutrons that make up a nucleus are held together by strong, attractive nuclear forces. Consider an α-unstable nucleus of atomic number Z. We may suppose that there is a special clumping that preforms the incipient α particle that is going to be ejected. Once the α particle is emitted and gets beyond the effective range of the attractive *nuclear* force exerted by the daughter nucleus, it experiences only the long-range Coulomb potential $V(r) = +2(Z - 2)e^2/r$. The factors here reflect that the α particle has charge $2e$; the daughter nucleus, charge $(Z - 2)e$. Thus, the α particle experiences a strong attractive force when it is inside the nucleus (whose radius R is roughly 10^{-12} cm) and a repulsive electrostatic force farther out. A cartoon version of this potential is shown in Fig. 5.1. The potential achieves its maximum value, V_{max}, at the nuclear radius. Let E be the energy of the α particle. Except perhaps for the very shortest-lived nuclei, this energy is usually well below the barrier height; that is, $E < V_{max}$. For example, in the case of the uranium nucleus U^{238}, E is about 4 MeV, the barrier height about 30 MeV. Classically, therefore, the α particle would not be able to escape the grip of the nucleus. Quantum mechanics, however, lets it tunnel its way out through the barrier. The ease with which it can do this depends very sensitively on V_{max} and on the α particle energy E. This explains why lifetimes of α-unstable nuclei vary over such a wide

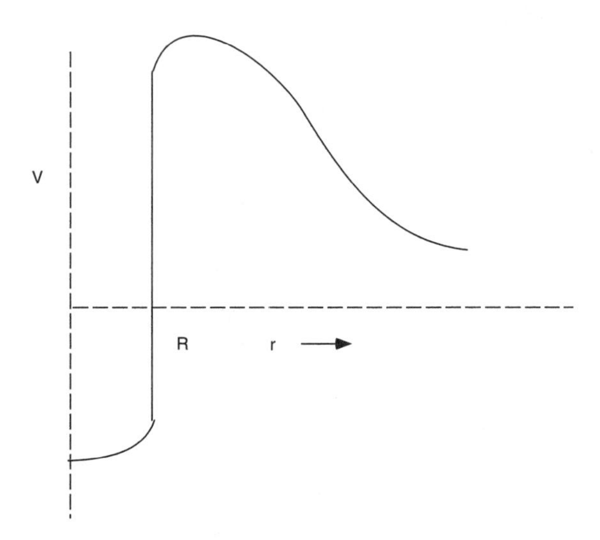

Figure 5.1 A cartoon depiction of the potential felt by an alpha particle that has formed inside a nucleus. The potential is strongly attractive within the nuclear radius R. Outside, the alpha particle feels a repulsive Coulomb potential.

range, depending sensitively as they do on quantities that vary from one unstable nucleus to another.

There is a final remark to be made. We have spoken as if the α particle emerges with a precisely defined energy E; that is, as if it is in a definite energy eigenstate. That's not quite the case. Strictly speaking, it is in a state that is a superposition of continuum energy eigenstates. However, in typical nuclear cases the (root mean square) spread of energies is small. That spread is related to the mean lifetime τ of the parent nucleus by the "uncertainty" relation discussed earlier between energy and time; roughly, $\Delta E \approx \hbar/\tau$. This connection between lifetime and energy spread of the decay products is general. It holds for any decay process. The energy spread is scarcely noticeable in many cases. It comes to about 6.6×10^{-16} eV if the lifetime is 1 second! Imagine then how small it is, say, for U^{238}, with its lifetime of several billions of years. But for certain subnuclear decay processes the lifetimes are sufficiently short to produce a noticeable energy spread. Indeed, for the very shortest-lived particles one determines the lifetime not directly but from the measured energy spread.

CHAPTER SIX

Identical Particles

Symmetry, Antisymmetry Rules

Although some of the principles of quantum mechanics were laid out earlier in general terms, for the most part we have concentrated so far on the case of a single particle. As the number of particles in a quantum system increases, the computational complications inevitably increase—often beyond reach if one is hoping for exact answers. Models based on physical insight and reasonable mathematical approximations have to intervene. However, as long as the particles in a system are all different one from another no new principles peculiar to multiparticle systems come into play. But, remarkably, the various elementary particles of nature *do* in fact come in strictly identical copies. Why this is so is something we'll turn to later on. For the present, let us just see how quantum mechanics deals with particle identity.

Both classically and quantum mechanically, two particles are said to be identical if they respond in exactly the same way to *all conceivable probes*. If the probe is a mass scale, they display the same mass; if it is an electric or magnetic field, they reveal the same charge; they scatter light waves in the same way; and so on. Classically, of course, if the objects are macroscopic you can mark and thereby distinguish them with identifying tags. But that's cheating: the marked objects are no longer identical. We are concerned here with strictly identical entities that cannot be tagged. Anyhow, classically there is no need to physically mark the particles. Although they are intrinsically identical, you can in principle keep an eye on them and simply declare at some initial instant that particle 1 is the one that's here, particle 2 is the one over there, and so on. Thereafter you can (in principle) follow their movements and thereby maintain a consistent identification. Whatever the force field in which the particles are moving, it treats them in an identical fashion, as is implied by the very hypothesis that the particles are identical. But the initial conditions were not identical (particle 1 was here, 2 there); therefore their orbits are different; and thus you know which particle is where. Classically, therefore, no special principles have to be invoked if one is dealing with identical particles.

The quantum mechanical situation is very different, since one deals not with definite particle locations, but only with probabilities. It may happen that the initial two-particle wave function is peaked such that the joint probability distribution is concentrated with one particle in the vicinity around here, the other in the vicinity around there. The coordinates for "here" you may ascribe to particle 1, for "there" to particle 2. But this distinction can get washed out over time, since wave packets move and change shape as time goes on. What were distinct peaks initially may broaden and overlap.

The way that quantum mechanics deals with particle identity has to be, and is, very different; and it has far-reaching consequences. According to the very meaning of identity, the Hamiltonian (energy) operator governing a system of identical particles will of course involve them on a totally symmetrical footing. It will be symmetric under the combined interchange of the position and spin labels of any one particle with those of any other. For example, if the labels $\mathbf{r}_1$ and $\mathbf{S}_1$ correspond to the coordinate and spin observables of particle 1 and $\mathbf{r}_2$ and $\mathbf{S}_2$ to the observables of particle 2, then the Hamiltonian will be symmetric under the joint interchange of $\mathbf{r}_1$ and $\mathbf{r}_2$ together with $\mathbf{S}_1$ and $\mathbf{S}_2$, though not necessarily symmetric if only the position or only the spin observables are interchanged. Similarly for the interchange of any other pair of identical particle labels in the system.

The wave function of a system of N identical particles depends on the coordinate variables $\mathbf{r}_i$ and spin quantum number "variables" m_{s_i}, $i = 1, 2, \ldots, N$. To avoid a surfeit of symbols, let us denote these variables using a compact notation for the arguments of the wave function. For any state Ψ, indicate the variables by writing $\Psi(1, 2, \ldots, N, t)$, where the number 1 stands for both $\mathbf{r}_1$ and m_{s_1}, the number 2 for both $\mathbf{r}_2$ and m_{s_2}, and so on. On purely mathematical grounds, the wave function of an identical particle system needn't have any special symmetry properties, even though the Hamiltonian equation that governs its time evolution is symmetric, as described above. However, nature is more demanding. There are quantum mechanical rules, and they are as follows.

> The wave function of a system of identical particles whose spin quantum number s is an integer must be totally *symmetric*. Particles of integer spin s are called *bosons* (named after the Indian physicist Satendra Bose).
>
> The wave function of a system of identical particles whose spin quantum number s is a half-odd integer must be totally *antisymmetric*. Particles of half-odd integer spin s are called *fermions* (named after the Italian-American physicist Enrico Fermi).

As explained above, symmetry—and now also antisymmetry—is in reference to the behavior of the wave function under the permutation of any two identical particles (the joint interchange of their coordinate and spin labels). If the wave function is symmetric, it is even (unchanged) under this permutation; if antisymmetric, it is odd (it changes sign). One can easily check that the above quantum mechanical rules are robust in the following sense. If the system wave function is symmetric at any one instant, that property will be maintained over time. This is so thanks to the symmetry of the Hamiltonian governing the time development of the wave function. So too, if the wave function is antisymmetric at one instant, that property will be maintained over time. Notice too that despite initial appearances, the *antisymmetry* property of fermion wave functions does not imply a lack of *symmetry* of physical phenomena. The probability of any physical occurrence will inevitably involve a product of the wave function and its complex conjugate. Since for fermions both change sign under a permutation, the probability amplitude does not; that is, it is symmetric.

These boson and fermion rules had to be taken as empirical discoveries at the time of their formulation in the period around the birth of nonrelativistic quantum mechanics. But soon they were seen to emerge as necessary consequences arising from the general ideas of relativistic quantum field theory. The "elementary" particles of everyday life—electrons, protons and neutrons—are spin-1/2 fermions. Photons, the other ingredient of everyday life, are spin-1 bosons. But what about composite particles; for example, nuclei? The answer here is that in the context of phenomena in which changes in the internal structure of nuclei do not come into play, as in much of chemistry, biology, materials science, and the like, nuclei can be treated as elementary particles obeying the appropriate symmetry or antisymmetry rules. For example, the nucleus of helium-4 is composed of four fermions (two protons, two neutrons). The interchange of two helium nuclei is therefore equivalent to the interchange of four pairs of fermions. For each pair there is a minus sign; altogether, therefore, a plus sign. Thus, the helium nucleus is a boson. More generally, nuclei with an even number of neutrons plus protons are bosons. They are fermions if the number of protons plus neutrons is odd. However, an important qualification is needed here. Typically, there are many different internal energy states of nuclei, just as for atoms. The notion of identity applies only to nuclei occupying the same internal states. For example, two carbon (C^{12}) nuclei in the *same* ground state are identical, as are two nuclei in the *same* excited state. But they are not identical if, say, one nucleus is in the ground state, the other in an excited state. At everyday temperatures the nuclei of any given species that we usually encounter are all in the ground level; if that level is not degenerate, the nuclei are identical.

Here is an immediate curiosity that follows from the quantum mechanical stance on particle identity. Consider a reaction in which two electrons come into collision, scatter, and reemerge moving in changed directions. Suppose that the electrons approach each other with equal and opposite momenta (equal in magnitude but oppositely directed), so that the total initial momentum is zero. By momentum conservation, the total momentum remains zero after the collision; hence the scattered electrons again have equal and opposite momenta. Let θ be the angle of scattering. We will be concerned here with the distribution function $P(\theta)$ that describes the probability distribution for scattering through angle θ. That distribution can be obtained, conceptually, by repeating the experiment over and over again with closely spaced detectors placed in all the scattering directions. Realistically, however, instead of repeated experiments with a single pair of colliding electrons one employs colliding *beams* of electrons. Now suppose (as is actually the case to a good enough approximation) that we can neglect spin-dependent forces and that the mutual potential energy of any pair of electrons is central (it is, in fact, the familiar Coulomb potential, but we can afford to be more general here). Consider two different situations.

(1) Antiparallel spins: the two incoming electrons have their spins pointing in opposite directions along some prescribed but otherwise irrelevant axis. For example, the electron coming in from the left has spin up; the one from the right, spin down.

(2) Parallel spins: the spins are both up (or both down) along the same direction, whatever that direction may be.

Assume that the detectors count scattered electrons without regard to spin direction. Then, since the electron-electron interaction force is by hypothesis spin independent, one might expect that the angular distribution function $P(\theta)$ will be the same for the two cases described above. But the distributions are in fact not the same. The reason is as follows. In case 2, since the spins are parallel the spin part of the system wave function is obviously symmetric. But the overall wave function must be antisymmetric. The spatial function must therefore be antisymmetric. In case 1, the wave function is a linear combination of two terms. One of them is the same as in case 2, but the other has a spatially symmetric wave function accompanying an antisymmetric spin part. Overall, therefore, the spatial functions for case 1 and case 2 are different. Thus, although the force law and detectors are spin insensitive, the requirement of combined space-spin antisymmetry leads to spin-dependent effects. The situation is especially dramatic at $\theta = \pi/2$ (90°). In case 1, $P(\pi/2)$ has some nonzero value. For case 2, the distribution function must precisely vanish, $P(\pi/2) = 0$.

The Pauli Principle

The fermion rule is sometimes presented under the name of Wolfgang Pauli, in the following manner: two identical fermions cannot be in the same state. This phrasing is loose, however, since for a multiparticle system there is in general no proper notion of *individual* particle states. The system wave function involves all the particles jointly. But there are special circumstances in which the states of interest *are* built up out of one-particle states. Let v_n be a set of one-particle states, that is, functions of the position **r** and spin quantum number m_s of a single fermion. The one-particle states we have distinguished by a counting index n. For a system of two identical fermions there is a special class of two-particle states that involve the antisymmetrized product of these one-particle states:

$$u_{n,n'}(1,2) = \frac{1}{\sqrt{2}}\{v_n(1)v_{n'}(2) - v_n(2)v_{n'}(1)\}.$$

The arguments 1 and 2 refer to the position and spin quantum number variables of particles 1 and 2, respectively. Clearly, $u_{n,n'}$ is antisymmetric under the interchange of the arguments 1 and 2. Referring to this kind of two-particle wave function, one may properly say that one of the two particles is in the one-particle state v_n, the other in the one-particle state $v_{n'}$. But you cannot go on to specify which particle, 1 or 2, is in which of these states. They take turns, so to speak, in the above formula. Moreover, there is clearly no two-particle state with $n = n'$, that is, with both electrons in the same one-particle state. That is the Pauli principle at work in this context. What has been described here for two particles can be generalized to a system of any number N of fermions. Take a product of N one-particle states $v_n(1)v_{n'}(2)v_{n''}(3)\ldots$, then antisymmetrize to form the N-particle state $u_{n,n',n'',\ldots}$. Of course the labels n, n', n'', etc. must all be different. This multiparticle function is said to have one of the electrons in the (one-particle) state v_n, another in the state $v_{n'}$, another in $v_{n''}$, and so on. Again, it makes no sense to say which electron is in which one-particle state. They take turns. For the class of multiparticle states described here, it now has meaning to say that two (identical) fermions can never be in the same (one-particle) state. Antisymmetrization wipes out that possibility.

This special class of multiparticle functions would seem to be of limited interest. In the following sense, however, it is not. Let v_n $(n = 1, 2, \ldots)$ be a complete set of one-particle states. What completeness means is that an arbitrary one-particle function is expressible as a linear combination of the v_n. It then follows that an arbitrary (antisymmetric) multiparticle function can be expressed as a superposition (over different choices of the set $n, n', n'', \ldots$) of the special multiparticle functions discussed above. To be sure, although this may be of mathematical interest, it may or may not be comforting when one is faced with some particular quantum mechanical problem, for example, finding the energy eigenvalues for a system of identical particles. The question is, under what conditions do we encounter these special antisymmetrized product states not in superpositions but taken individually?

In the important context of the energy eigenvalue problem for a system of N identical fermions, we encounter the necessary conditions if, whether exactly or in some reasonable approximation, we can ignore interparticle forces within the system, so that the only forces are external ones. In that case, it turns out that the multiparticle eigenvalue problem is reduced to solving the one-particle problem. Take the states v_n to be the energy eigenfunctions for a single particle moving in the external force field and let ε_n be the corresponding energies. Here again, n is a counting index that distinguishes one state from another. It is now very easy to see that the eigenfunctions of the N-particle problem are just the antisymmetrized products $u_{n,n',n'',\ldots}$ that we discussed above. The eigenfunctions are specified by the set of N one-particle labels, $n, n', n'', \ldots$. The corresponding energies are just the sums of one-particle energies:

$$E_{n,n',n'',\ldots} = \varepsilon_n + \varepsilon_{n'} + \varepsilon_{n''} + \cdots. \tag{6.1}$$

This result agrees with one's intuition; namely, that the multiparticle energies are additive in the one-particle energies since by assumption there are no interparticle forces.

It is now time for examples.

The Fermi Gas

The outer atomic ("valence") electrons in metals are not bound to individual atoms but instead move more or less freely in the metal sample with energies that lie in so-called "conduction bands." The positively charged ions essentially remain in place, forming a regular array and undergoing small vibrations about their average positions in the array. Realistically, of course, the conduction band electrons interact among themselves as well as with the positive ions. It would be wrong to ignore these interparticle forces. But let us do that anyhow, and indulge in a highly oversimplified treatment embodied in the so-called free electron model. Surprisingly, it is not hopelessly wrong; it captures at least some of the phenomenology. And it is simple.

One Dimension

We will start in one dimension for warm-up. Consider a system of N identical spin-1/2 fermions (call them electrons), moving freely in a one-dimensional box with walls at $x = 0$ and $x = L$. We shall suppose that N and L are both macroscopic (that is, very large). The ratio N/L is the average number density (the number of electrons per unit one-dimensional "volume"). Let us compute the ground state energy of this system. According to the discussion leading to Eq. (6.1), since we are assuming that the electrons do not interact among themselves or with the ions, it is enough to solve the energy eigenvalue problem for a single free fermion in the box. We have already done that, with the result given in Eq. (5.3)—except that the energies written there will now be denoted by the Greek letter ε, the Roman E being reserved for the N-particle system. The ground state of the latter is obtained by putting 2 electrons (spin up, spin down) in the one-particle spatial state $n = 1$, 2 electrons in $n = 2$, and so on, until all N electrons have been spoken for. Owing to the Pauli principle, there can be no more than 2 electrons in each one-particle spatial state. For simplicity, take N to be even, so that $n_{\max}$, the maximum n value of any occupied state, is $n_{\max} = N/2$. (If N happens to be odd we will be off by very little at the end if N is also very large.) The ground state energy of the whole system is thus

$$E_{\text{gnd}} = 2\frac{\hbar^2\pi^2}{2mL^2}\left\{1 + 2^2 + 3^2 + \cdots + \left(\frac{N}{2}\right)^2\right\}.$$

The factor 2 in front on the right hand side is the number of spin states for each value of the spatial state index n. For large N, to within a correction of order $1/N$, the sum can be replaced by an integral and is then easily evaluated. The ground level energy, per particle, turns out to be

$$\frac{E_{\text{gnd}}}{N} = \frac{\hbar^2\pi^2}{24m}\left(\frac{N}{L}\right)^2. \tag{6.2}$$

Notice that the energy per particle depends on N and L only through their ratio; that is, it depends only on the number *density* N/L.

Three Dimensions

Now take the case of N free electrons in a cubical box of side L. We will again be interested in the macroscopic limit where the number of particles N and the volume L^3 are both large; N/L^3 being the average number density. Again, let us compute the ground level energy. The procedure is exactly as in the one-dimensional case, but now we refer to Eq. (5.4) for the one-particle energy levels. Each such spatial level is labeled by a set of three integers, n_1, n_2, n_3. The lowest one-particle energy corresponds to $(n_1, n_2, n_3) = (1, 1, 1)$. Place two electrons in that spatial state, spin up and spin down. Next come the degenerate one-particle states $(1, 1, 2)$, $(1, 2, 1)$, and $(2, 1, 1)$. Place two electrons in each. And so on, moving on up to higher and higher one particle levels until all N electrons are accounted for. The N-particle ground state energy is then just a sum of the one-particle energies. For large N, to within a negligible correction of order $1/N$, the sum can be replaced by an integral. The ground level energy, per particle, then turns out to be

$$\frac{E_{\text{gnd}}}{N} = \frac{3}{5}\varepsilon_f, \quad \varepsilon_f = \frac{\hbar^2}{2m}\left(3\pi^2\frac{N}{V}\right)^{2/3}, \quad V = L^3. \tag{6.3}$$

The so-called *fermi energy* ε_f is the energy of the highest one-particle state that is "occupied" when the multiparticle system is in its ground level. It depends on the total number of electrons N and total volume V only in their ratio, the number density N/V, and it varies as the two-thirds power of that density. The average electron energy, E_{gnd}/N, is equal to three-fifths of the fermi energy.

The workings of the Pauli principle are evidently quite dramatic. If there were no Pauli restriction, the ground state of the N-particle system would have all of its electrons in the lowest one-particle spatial state. In that case E_{gnd}/N would be proportional to $1/L^2$, which is essentially equal to zero for macroscopic L. The Pauli principle instead distributes the electrons over a range of one-particle energies on up to the fermi energy. With the electron number densities encountered in the conduction bands of real metals, fermi energies are typically of order a few to ten or more electron volts. For many purposes, energies in this range can be regarded as large: large, that is, compared to characteristic thermal energies $k_B T$. Here k_B is Boltzmann's constant and, as always, T is temperature on the absolute scale (absolute zero occurs at -273° on the Centigrade scale). It is convenient to define a fermi temperature T_f according to

$$k_B T_f = \varepsilon_f.$$

Characteristic fermi temperatures range from several tens of thousands to a hundred thousand or so degrees Kelvin! Thus, for metals at anything like realistic temperatures, $T \ll T_f$.

Even within the crudities of the free electron model, in order to understand the role of conduction band electrons in metals, one of course has to deal not just with the ground state but also with excited states. Any multiparticle state is characterized by telling which of the one-particle states are occupied. For the multiparticle ground level, all of the one-particle states with energies up to the fermi energy are occupied, but none above. In the various excited levels of the many-body system, some of the electrons are in one-particle states above the fermi level. This is of course accompanied by a corresponding depletion below the fermi level (the depleted one-particle states below the fermi level are often referred to as "holes"). At any finite temperature a fermi gas of electrons will be in a mixture of energy eigenstates. At normal temperatures this mixture is dominated by the ground state as well as low-lying excited states in which only a small fraction of the electrons are above the fermi energy—not far above. Under normal conditions, therefore, it is only the electrons right around the fermi level that do the electronic work of a metal, for example, in connection with heat and electric current conductivity. This is so because electrons well below the fermi level cannot easily absorb or yield the small increments of energy involved in phenomena at ordinary temperatures: the nearby one-particle states above and below them are for the most part already occupied, and Pauli does not allow multiple occupancy.

A striking property of the fermi gas is that even at very low temperatures—indeed, even at the absolute zero of temperature—it exerts a pressure. Let us consider this limiting temperature. For $T = 0$ we can take the system to be in the ground level; and as is evident from Eq. (6.3) the energy of that level is a function of the volume V. The smaller the volume the greater the energy. To compress the gas one therefore has to supply energy by exerting a force, say on one of the walls acting as the face of a piston. That bespeaks a pressure exerted by the gas on the walls. The pressure P is in fact the negative derivative of energy with respect to volume. Working this out, one finds that the product of pressure and volume is given by

$$\text{fermi gas:} \qquad PV = \frac{2}{5} N \varepsilon_f. \tag{6.4}$$

For comparison, the classical, *ideal gas law* that one learns in school is

$$\text{ideal gas:} \qquad PV = N k_B T. \tag{6.4$'$}$$

At $T = 0$ the ideal gas exerts no pressure. The quantum fermi gas does. For temperatures large compared to the fermi temperature, the fermi gas equation of state reduces to that of the classical idea gas. It is in the domain $T \ll T_f$ where the quantum system departs so dramatically from the classical. In that domain the fermi gas is said to be *degenerate*, and one speaks there of *degeneracy pressure*.

The conduction band electrons in metals are well within the degeneracy regime, and degeneracy pressure is an important contributor to the bulk modulus of metals (bulk modulus relates pressure change to corresponding volume change). Degeneracy pressure also plays a role in the heavens. A normal star such as our own sun is composed mainly of electrons and ionized hydrogen and helium. The hydrogen is burning, converting to helium, positrons, and neutrinos through a series of nuclear rather than chemical reactions. The electrons and other entities are essentially in an ideal gas regime in which temperatures and densities adjust themselves so that the gas pressure stabilizes the star against gravitational collapse. Gravitational collapse is of course a threat because the gravitational force is attractive; it wants to pull together the parcels of matter. The gas pressure resists this. As the hydrogen burns away the star begins to collapse gravitationally. This implies increasing density and hence increasing fermi temperature; the electron gas will eventually enter the degeneracy regime. If the star is not too massive, so that the gravitational forces are not too strong, the electron degeneracy pressure will suffice to once again stabilize the star, now in a new incarnation as a white dwarf. The limiting mass, first worked out by S. Chandrasekhar, is about 1.4 solar masses. In the white dwarf stage the star is no longer burning; but it starts out very hot, thanks to the energy extracted from the gravitational collapse that carried it to that stage. It then cools over the subsequent eons. A typical white dwarf density is about 10^7 times that of the sun. Its radius is roughly that of the earth. The central temperature is of order 10^7 degrees Centigrade. That seems a lot, but it is as nothing compared to the fermi temperature, which is about 10^{11} degrees Centigrade. So far as the electrons are concerned, the white dwarf is at the absolute zero of temperature.

If a star is too massive to be saved from gravitational collapse by the electron gas, it will pass to a condition of such high density that it converts to a system of neutrons, the electrons and protons having largely disappeared through the reaction electron + proton $\rightarrow$ neutron + neutrino, with the neutrinos escaping the star altogether. If the star is not *too* massive, *neutron* degeneracy pressure may provide a successful second-ditch stand of stability against gravitational collapse. In that case the star ends up as a neutron star, or *pulsar*. The analysis here is more tricky than for white dwarfs because neutron-neutron interactions are so strong that it is not realistic to treat the neutron system as a gas of noninteracting fermions. In any case, if the star is too massive to be saved even by the neutrons, and if it cannot shed enough of its excess mass in a supernova explosion, it will go on to collapse to a black hole. Quantum mechanics at the black hole level is a lively and open subject of contemporary research.

Atoms

The one-electron atom was easy. For many-electron atoms, exact analytic solutions of the energy eigenvalue equation are beyond reach. Indeed, with increasing number of electrons a precision all-out *numerical* attack becomes hopelessly demanding even for modern computers. But the experts in this well-developed field have successfully devised various approximation procedures based on plausible physical models (which still require substantial globs of numerical computation). The virtue of modeling, if it is good modeling, is that it nourishes physical intuition and provides a useful basis for organizing, interpreting, and communicating numerical results. In the following discussion we ignore spin-dependent forces and relativistic corrections. The multielectron atom is already complicated enough.

If one could ignore the forces that the electrons exert on one another, so that the electrons could be treated as moving independently in the attractive potential of the nucleus, matters would be easy. The multiparticle eigenstates would be antisymmetrized products of one-particle eigenstates; the corresponding energies would be sums of one-particle energies—all of this as discussed above. It would therefore be enough to solve the one-particle problem. Moreover, for the Coulomb potential we of course already know the solutions analytically. The trouble is that it is *not* a good idea to ignore the electron-electron interactions in an atom. To get a feel for that, let's consider the ground state of the two-electron atom. For a *single* electron in the field of a nucleus of atomic number Z, the energy levels are given by Eq. (5.15). In what follows, we replace the symbol E_n there by the symbol ε_n, to emphasize that this is a one-particle energy. Numerically, the one-particle energies are

$$\varepsilon_n = -13.6\frac{Z^2}{n^2} \text{ electron volts.}$$

For the two-electron atom, with the electron-electron potential ignored, the ground state would have both electrons in the $n = 1$ spatial state—one electron with spin up, the other with spin down. For helium ($Z = 2$), the predicted ground state energy would therefore be -108.8 eV. But the experimental value is -78.9 eV. The discrepancy here is substantial. It is simply not a good idea to ignore the electron-electron interaction. And so too for atoms with more than two electrons.

It is necessary, therefore, to look for approaches that incorporate these interactions in some reasonable approximation and that are at the same time computationally tractable. The nature of the approximations adopted will depend in part on the kinds of questions one is tackling (for example, whether they concern the ground and low-lying states or highly excited states of the atom); also, on what one accepts as "tractable." Especially for ground and low-lying states, an approach whose objective is at least easy to describe, if not so easy to implement computationally, is based on the following. In a multielectron atom, any one electron is acted on by all of the others as well as by the nucleus. If one knew the spatial probability distribution of the other electrons, it would be possible to compute the net force exerted on the electron under consideration by its fellows and, of course, by the nucleus. By this means one could compute an *effective potential* acting on the electron, a potential that takes into account the other electrons. But this probability distribution is not known until the multiparticle energy eigenvalue problem has been solved, so this looks to be going in circles. However, various approximation procedures have been devised by which one can make a trial guess of an effective potential and then improve on it self-consistently, or in other ways come to a plausible choice of effective potential. One then treats the electrons as if they move independently in this potential, assured (putting the matter optimistically) that the electron-electron forces have been taken into account, at least approximately. Usually one goes a bit farther and restricts oneself to finding a plausible *central* potential.

The procedures for coming up with an effective potential are highly technical. Just to drop the names of two of the more popular approaches, there is the *Hartree-Fock* approximation and the *Fermi-Thomas* model. Once an effective central potential V_{eff} has been settled on, one way or another, the solution of the multiparticle bound state problem reduces to solving the one-particle problem in that potential. This is because in the approximation under discussion the electrons are to be treated as moving independently in the effective potential. That potential, of course, no longer obeys the Coulomb $1/r$ law. It is bound to be more complicated, and the one-particle problem can no longer be solved analytically. But here modern computers can cope well enough. One breathes a sigh of relief once the multiparticle problem has been reduced to a one-particle task. The really hard thinking and computational work goes into getting a good effective potential. Bear in mind that that potential is in no sense universal, it is different for different atoms (different numbers of electrons).

Suppose we're dealing with the effective potential for the ground state (or some low-lying excited state) of a neutral atom containing Z electrons. We can anticipate some limiting properties that a reasonable effective potential $V_{\text{eff}}(r)$ ought to have.

(1) As an electron moves very close to the nucleus, the unshielded nuclear Coulomb potential should dominate. We therefore expect that

$$V_{\text{eff}}(r) \to -Ze^2/r, \qquad \text{as } r \to 0.$$

(2) As an electron moves very far away from its fellow electrons and the nucleus, it sees a small blob of stuff (the nucleus shielded by the remaining $Z-1$ electrons) of net charge e; so we expect that

$$V_{\text{eff}}(r) \to -e^2/r, \qquad \text{as } r \to \infty.$$

For distances neither very small nor very large, the potential function is bound to be complicated.

Whatever the details of that potential, since it is central (by construction), we know that the one-particle energy observable commutes with the orbital angular momentum observables L^2 and L_z; also with the electron spin variable S_z (see the discussion of central potentials in Chapter 5). The one-particle energy eigenstates are thus labeled by the orbital angular momentum quantum numbers ℓ and m_ℓ, the spin quantum number m_s, and a principal quantum number n. The corresponding one-particle energies $\varepsilon_{n,\ell}$ depend only on n and ℓ. Recall that the degeneracy is $2(2\ell + 1)$, where the factor 2 in front comes from the fact that m_s can take on only the two values $1/2$ (spin up along the z axis) and $-1/2$ (spin down); the factor $(2\ell + 1)$ is the number of possible values of the quantum number m_ℓ. For given ℓ, m_ℓ, and m_s, the principal quantum number is a counting index that distinguishes states of different energy. By convention, for given ℓ the numbering begins at $n_{\text{min}} = \ell + 1$.

This is a good place to introduce a notation that has long been conventional, first in atomic physics and then in wider contexts. Associate to each ℓ value a letter of the alphabet, according to the conventions:

ℓ value	letter designation
0	s
1	p
2	d
3	f
.	.
.	.
.	.

The listing is alphabetical beyond the letter f. The letter e is omitted altogether, for fear of confusion with electric charge. Of course one eventually runs out of letters and thereafter reverts back to a numerical designation of ℓ. But so long as letters are in use for ℓ, the principal quantum number n and the letter designation for ℓ are joined together in expressions such as $2s$, $4p$, and so on, which, respectively, denote the one-particle states $(n = 2, \ell = 0)$, $(n = 4, \ell = 1)$, and so on. You should never encounter a state such as $3f$, since this violates the counting convention that n cannot be smaller than $\ell + 1$.

By that very counting convention, the energy $\varepsilon_{n,\ell}$ increases with increasing n for given ℓ. The degeneracy in ℓ that held for the Coulomb potential of course no longer obtains. We will not record here the actual numerical values of the one-particle energies. These vary in any case from one atom to another owing to the fact that the effective potential is different for different atoms. Anyhow, there are different effective potentials on the market. However, we can at least give some indication of the ordering of levels on the energy scale. For typical atoms the sequence, starting with the lowest energy, is $1s$, $2s$, $2p$, $3s$, $3p$, $\{4s, 3d\}$, $4p$, $\{5s, 4d\}$, $5p$, $6s$, $\{5d, 4f\}$, $6p$, $7s$, $\{6d, 5f\}$. We won't need to go beyond this, even for uranium. Curly brackets have been placed around levels that are close in energy, their relative order being reversed in some atoms. It should be noted here that for given ℓ the spatial wave function is increasingly spread out in the radial direction as the principle quantum number increases; that is, the mean radius $\langle r \rangle$ grows with n. It is also the case, reflecting a kind of centrifugal repulsion, that the radial wave function is increasingly suppressed near the origin, that is, near the nucleus, as the angular momentum quantum number ℓ increases for given n. Finally, before we go on we should recall that this whole effective potential procedure is an approximation, designed, insofar as we are discussing it here, to deal with the ground states of multielectron atoms and, less securely, the low-lying excited states.

We are now ready for the atoms. In the approximation under discussion, any state of a multielectron atom is fully specified by listing the one-particle states that are occupied. Recall that the latter are specified by the four quantum numbers n, ℓ, m_ℓ and m_s, the energies $\varepsilon_{n,\ell}$ depending only on n and ℓ. Thanks to Pauli, the occupation number for any one-particle state can be only 0 or 1. However, although two electrons cannot share all four quantum numbers, two or more electrons can share the quantum numbers n and ℓ and thus share the same one-particle energy, provided they are in states differing in one or both of

m_ℓ and m_s. The set of $2(2\ell+1)$ one-particle states differing in m_ℓ and m_s but sharing the same n and ℓ quantum numbers constitute what is called a *shell*. Thus, an ns shell can accommodate up to 2 electrons, an np shell up to 6, an nd shell up to 10, and so on.

Start with helium. The ground level clearly must have both electrons in a $1s$ state, one with spin up, the other with spin down. One speaks of this as a $(1s)^2$ *configuration*. The $1s$ shell is said to be filled, or closed. The helium ground state is tightly bound. The empirical *ionization energy*, the energy needed to remove one electron while leaving the remaining electron in the ionic ground state, is $I = 24.6$ eV. That's a lot. That is why the helium atom is so chemically unreactive. It is in fact hard to drag an electron even part way out to help bond the helium atom to other atoms. Helium is an inert gas.

The neutral lithium atom has three electrons. The $1s$ shell cannot accommodate all three, so the ground level of lithium has the configuration $(1s)^2(2s)$: 2 electrons in the $1s$ shell, another in the $2s$ shell. If electron-electron interactions could be ignored, the energy needed to remove the $2s$ electron would be 30.6 eV, a value that comes from the hydrogenic formula $13.6(Z^2/n^2)$ eV with $Z = 3$, $n = 2$. At the other extreme, if it could be assumed that the nuclear potential is maximally shielded by the two $1s$ electrons, the $2s$ electron would see an effective atomic number $Z^* = 1$. The ionization energy would be only 3.4 eV. The empirical ionization energy is in fact $I = 5.4$ eV. It is as if the effective charge parameter seen by the *valence* electron outside the closed $(1s)^2$ shell is $Z^* = 1.3$. That corresponds to a lot of shielding, but not quite maximal.

The beryllium atom has its four electrons in the configuration $(1s)^2(2s)^2$. This is again a completely closed shell configuration, as for helium. Unlike helium, however, beryllium is not chemically unreactive. That's because the $2p$ level happens to be only slightly higher in energy than the $2s$ level. Other atoms that bond with beryllium exploit this by supplying the small amount of energy needed to promote a beryllium electron from $2s$ to $2p$, gaining energy in return by rearranging their own electronic structures in such a way as to bond. The details of chemical bonding are beyond the reach of our present lightning survey of atoms. Starting with boron ($Z = 5$) and passing on through carbon (6), nitrogen (7), oxygen (8), fluorine (9), and on to neon (10), one adds each new electron to the $2p$ shell so that boron has the configuration $(1s)^2(2s)^2(2p)$; carbon, the configuration $(1s)^2(2s)^2(2p)^2$; on up to fluorine, $(1s)^2(2s)^2(2p)^5$; and neon, $(1s)^2(2s)^2(2p)^6$.

Neon is an inert gas. Its shells are all closed. A considerable amount of energy would be needed to pull any of the electrons into a bonding arrangement. Fluorine is one electron shy of being a closed $2p$ shell configuration. That makes it hungry for an extra electron. It is chemically reactive, eager to accept an electron from a bonding partner. Sodium ($Z = 11$) has one electron beyond neon, and that extra electron has to go into the $3s$ shell. That is, sodium has the configuration (Ne)($3s$), where, as a space saver, (Ne) is written to symbolize the configuration of neon. Magnesium has the configuration (Ne)$(3s)^2$. This is a closed shell situation but, as with beryllium, magnesium is not inert owing to the fact that the $3p$ level is not much removed from the $3s$ level on the energy scale.

The next inert configuration occurs for argon ($Z = 18$), which has a filled p shell. The configuration is (Ne)$(3s)^2(3p)^6$. The long sequence from potassium to krypton builds on the argon configuration, adding first to the $4s$ shell, then to $3d$ (with a little bit of shuffling between these two competitors along the way), then to the $4p$ shell. Like helium, neon, and argon before it, krypton ($Z = 36$) is chemically inert. Its configuration is (Ar)$(4s)^2(3d)^{10}(4p)^6$. The sequence from rubidium to xenon builds on the krypton configuration, adding $5s$, then $4d$ electrons (with some shuffling back and forth), then $5p$ electrons. Xenon has the configuration (Kr)$(5s)^2(4d)^{10}(5p)^6$. And so it goes on. We'll stop our atomic journey here.

More on Identical Bosons

There is no Pauli principle for identical bosons. There is no limit, therefore, to the number of them that can occupy the same one-particle state. Indeed, in some respects they very much like to be together. Consider, for example, a free boson gas analogous to the free fermion gas discussed earlier; namely, a collection of N identical bosons occupying a macroscopic cubical box of volume L^3. If the box is large, the one-particle levels will be very closely spaced as judged on macroscopic standards. So far, there is no difference between bosons and fermions. How can there be, if we are talking about single-particle states? But for the multiboson gas, in contrast to the fermi gas, the ground state has all the bosons placed in the same, lowest, one-particle level. The energy of the N-boson ground state is therefore essentially zero if the system is macroscopic. But there is something else more striking. For a macroscopic boson system there is a whole spectrum of energy levels so closely spaced as to be practically continuous, extending upward from the ground level. At the absolute zero of temperature, the system must be in the multiparticle ground state; but at even tiny temperatures above zero, one would expect the system to be spread out over a whole range of low-lying N-particle levels. Indeed, there are so many of these, and they are so closely spaced, that no one of them—including the ground level—should carry much thermodynamic weight. Or so one might think. But it turns out to be otherwise! There is a celebrated thermodynamic phase transition, the so-called *Bose-Einstein condensation*, that is predicted for the free boson gas. It is as follows. There exists a certain critical temperature above which the expectation discussed above obtains: no substantial occupation of any particular one-particle level, including the one-particle ground level. But below that critical temperature a finite fraction of the bosons *condense*, as one says, into the one-particle ground state. We need not give the formula here for that critical temperature. It depends in a definite, calculable way on the mass of the boson and on the number density. The important thing is that this condensation effect shows up in the form of certain distinctive changes predicted for various thermodynamic properties, for example, specific heat, as the system crosses from just above the critical temperature to just below. Of course, the free boson gas is an idealization, but qualitative indications of the Bose-Einstein effect can be espied in certain real systems.

Of greater practical and equally fascinating scientific interest is a certain clumping tendency displayed by a species of bosons that we see every day—photons. The emission and absorption of photons by material systems, atoms for example, requires the machinery of quantum field theory for its proper understanding. Nevertheless, basing himself on the Planck blackbody spectrum and employing brilliant thermodynamic reasoning, Einstein could achieve a major insight as far back as 1917, in the days of the old quantum theory. He considered radiative transitions between any given pair of energy levels of an atom (or molecule). Let E_{I} and E_{II} denote the energies, and take $E_{\mathrm{II}} > E_{\mathrm{I}}$. In what follows, let us focus on photons of circular frequency $\omega = (E_{\mathrm{II}} - E_{\mathrm{I}})/\hbar$, moving in some specified direction with specified polarization. *Absorption* refers to a transition in which the atom jumps upward from level I to level II, absorbing an incident photon of the type in question. *Emission* refers to radiation of a photon as the atom jumps down from II to I.

It is intuitive (and correct) that the rate of absorption is proportional to the flux of incident photons. For emission, what Einstein deduced is that the rate is a sum of two terms, a *spontaneous emission* term and an *induced emission* term. The spontaneous term refers to emission that occurs even in the absence of preexistent photons in the vicinity. The induced term, just as for absorption, is a contribution that is proportional to the flux of preexistent photons of the given type. Thus, the more photons of that type that are already in the vicinity, the more eager the atom is to radiate yet another one. The photons in this sense like to be together. It is this phenomenon of induced emission that is at the heart of the laser. A cartoon description is as follows. Start with a system of atoms in the dark (so to speak), zap the system somehow to get ample spontaneous emission started, trap that radiation sufficiently well, and thereby build up the intensity through induced emission.

Superconductivity is another phenomenon in which the "togetherness" predilection of identical bosons is well exemplified.

Many metals, though not all of them, lose all electrical resistivity below some critical temperature. The transition temperatures T_C are very low, less than a few tens of degrees above absolute zero for the traditional "low-temperature" superconductors; not so low for a more recently discovered class of "high-temperature" superconductors, where T_C in some cases exceeds a hundred degrees above absolute zero. What is especially striking about superconductors apart from perfect electrical conductivity is their behavior in magnetic fields. If an external magnetic field is turned on after a metal has been cooled to a superconducting state, the field will not penetrate into the superconductor (proviso: the magnetic field must not be too strong). But suppose a magnetic field is established through a sample while it is still in a normal state. If the sample is now cooled to below the transition temperature, the magnetic field will be expelled from it. Now remove the *external* field source. In the space outside the superconductor there will still remain a magnetic field. It is produced by electrical currents that have been induced in the surface layers of the metal by the external field before it was removed. Because the superconductor has no resistance, that current, once induced, persists. Suppose that the sample is in the form of a ring. There will be a trapped magnetic flux passing through the area encircled by the ring. The magnitude of the trapped flux must, of course depend on the strength of the external magnetic field that was initially present, a strength which could a priori have had any value—it is a continuously adjustable parameter. The quantum mechanical surprise is that the trapped flux that remains after the external field is removed comes only in discrete units, multiples of a flux quantum $2\pi\hbar c/Q$, where $Q = 2e$ (e is the magnitude of the electron charge).

What has all this to do with identical bosons? Electric current in a metal is carried by moving electrons; and electrons are fermions, not bosons. But there is an interesting effect at work in superconductors (in what follows we are considering low-temperature superconductors). The Coulomb force between any pair of electrons is of course repulsive (like charges repel). But the electrons in a metal also interact with the positive ions that make up the metal's framework. The ions don't move much, but they do vibrate, each one about its equilibrium position. Through the intermediary of these vibrations, electrons exert a force on one another that goes beyond the direct Coulomb force. This comes about because any one electron exerts a force on the vibrational system, which in turn then exerts a force on another electron. In low-temperature superconductors, this force is attractive and outweighs the repulsive Coulomb force between any pair of electrons. The net result, very loosely speaking, is that electrons bind in pairs; and a bound system composed of two fermions is a boson with charge $Q = 2e$. Thus, the system of N conduction electrons in a superconductor can be thought of—very roughly—as involving a collection of such boson-like pairs. At low temperatures these bosons like to occupy the same state. In normal conductors, electrical resistance arises because the flowing electrons lose energy in collisions with the ions and with one another. In superconductors, the electrons bound in bosonic pairs are not easily disassociated.

CHAPTER SEVEN

What's Going On?

Quantum mechanics deals with probabilities. Observers deal with facts: meter readings, tracks in a photographic emulsion, clicks of a Geiger counter, and so on. The big question is, how do probabilities get converted into facts? The formulaic answer is that this conversion takes place whenever a measurement is performed on the quantum system under consideration. Operationally, as far as we know, that is the right answer; but it is very puzzling. The measuring apparatus on this view is regarded as lying outside the probabilistic structure of quantum mechanics. When called upon, it steps in and makes a definite selection from among the competing alternatives; and the system wave function "collapses" to the selected state. In a series of repeat measurements under identical initial conditions, the measuring apparatus will produce a range of different selections, the probability distribution being dictated by the rules of quantum mechanics. But for each individual measurement there emerges some particular outcome.

The trouble with this is that the apparatus A_1 used for a measurement is as much a part of nature as the quantum system Q_1 being interrogated. The two together constitute a larger quantum system Q_2, about which quantum mechanics once again makes only probabilistic assertions. Of course, if a new "outside" apparatus A_2 is introduced to perform measurements on Q_2, facts will once again emerge—a particular outcome each time. But in fact A_2 ought also to be part of nature, and we should then be able to regard $Q_2 + A_2$ as some larger quantum system Q_3, in which case we're back to probabilities only. And so on. There seems to be nothing within quantum mechanics that tells how to convert probabilities into facts.

Let us pursue this with an example. Suppose that the quantum system is a single particle with spin. To avoid certain complications that are irrelevant for the present discussion, take the particle to be electrically neutral; say a neutron (or neutral atom). Suppose that the observable of interest is the component of spin along some particular axis. Although the neutron is electrically neutral, it has a magnetic moment (as do many neutral atoms), and this provides a handle on spin. There is a standard setup first employed by O. Stern and W. Gerlach that involves an inhomogeneous magnetic field and serves to measure the spin component along any prescribed direction, say the z axis. It can do so because the inhomogeneous magnetic field exerts a force on the magnetic dipole; and the dipole moment is proportional to the spin angular momentum vector of the particle. In passing through the apparatus, the neutron's wave packet is bent in one direction (say to the right) if the spin is up, in the other direction (to the left) if the spin is down. Detectors are placed to the right and the left. If the detector on the right registers a count, one learns that the spin is up; if the detector on the left registers, the spin is down. We can imagine the detectors being hooked to a meter whose needle moves to a position labeled M^+ for spin up and to a distinctly different position M^- for spin down. Denote the neutral position of the needle by M^0. We may then characterize the situation as follows. Suppose that the meter is in the neutral position M^0 before the neutron enters the apparatus and that the neutron spin is up. Denote this initial state by the symbol $(\uparrow \cdot M^0)$, where $\uparrow$ is the symbol for spin up. Assuming that the neutron survives its passage through the detector with spin state unchanged, the state after the detection is $(\uparrow \cdot M^+)$: the needle is in the position M^+ and the neutron spin is up. The transition from pre- to post-measurement can thus be symbolized by

$$(\uparrow \cdot M^0) \rightarrow (\uparrow \cdot M^+). \tag{7.1}$$

If the spin is down, $\downarrow$, the measurement transition is

$$(\downarrow \cdot M^0) \rightarrow (\downarrow \cdot M^-). \tag{7.2}$$

It has to be said that there is quite a bit of idealization in this description. We are supposing that the measurement apparatus does its job perfectly. Inevitably, there will be imperfections. For example, the bending to the right or left that we spoke of refers to the center of mass of the neutron wave packet. Because to some extent that packet is spread out to start with and because it tends to spread additionally over time, it can happen that a packet bent to the right (left) has some small overlap with the detector on the left (right). However, this error can be made negligibly small in practice. Another idealization is this. We have treated the measurement apparatus as being characterized by the needle location; in this example, as if there are only three possible quantum states of the apparatus, M^+, M^0, and M^-. That is of course grossly wrong. The apparatus is a macroscopic system composed of an astronomical number of atoms. The space of states is enormous. However, we can imagine organizing these states into three very large families classified by the needle location observable. Mark off three finite, nonoverlapping intervals on the meter, corresponding to spin up, neutral meter setting, spin down. The whole class of states with the needle in the spin-up interval we have denoted collectively by the symbol M^+; and similarly for the other two intervals. If the apparatus is robustly designed and if the spin is up, the transition will go from *any* of the states in the family M^0 to some state in the family M^+ (but not to any states in the families M^0 and M^-); relatedly for spin down. That is, there are hordes of microscopic and even macroscopic variations to which the main result is insensitive; for example, the correlation of meter reading and spin is unaffected (within reason) by the temperature of the apparatus, by possible small cracks in the casing, by the logo imprinted on the magnet, and so on.

The extent to which the measurement setup actually displays the ideal behavior reflected in Eqs. (7.1) and (7.2) is something that can *in principle* be addressed purely within the framework of quantum mechanics, without reference to "outside" observers. Given a complete specification of the apparatus, one can *in principle* find all the relevant quantum states, organize them into the three classes described above, then solve the Schroedinger equation to check how well the results accord with Eqs. (7.1) and (7.2). In reality, of course, such a full-blooded quantum calculation is hopelessly out of the question. Sensibly, for the macroscopic apparatus that they design and use, experimentalists rely on a mixture of classical reasoning, good craftsmanship, and empiricism.

In connection with the spin measurement example, it should also be said that spin "up" and spin "down" do not fully characterize the state of the neutron. Its state is also a function of the position variable. Indeed, the Stern-Gerlach apparatus produces a correlation between space and spin that serves as a basis for determining the spin. The spin-up packet is bent to the right, the spin-down packet to the left. If you detect the packet bending to the right, for example, you have determined that the spin is up. That correlation is easy to prove theoretically, within quantum mechanics. But the question is, how do you in fact know in an experiment which way the packet is bending? Well, you ask which of the two detectors makes a response. But how do you know which one responds? Well, the position of the needle on the meter answers that. But how do you determine that position? Well, you arrange that if the needle is at M^+ a blue flash is emitted, if at M^- a red flash. But who detects the flash? And so on. Within its own formalism, quantum mechanics predicts correlations: if this, then that. But when there are competing possible outcomes of a measurement it does not in itself tell which of them actually materializes as a fact.

This situation is dramatized if we ask what happens when the incident neutron is in a state Ψ that is a superposition of spin up and spin down,

$$\Psi = a\uparrow + b\downarrow, \tag{7.3}$$

where a and b are constants, with normalization $|a|^2 + |b|^2 = 1$, where $|a|^2 = a^*a$, and so on. If the measurement apparatus obeys Eqs. (7.1) and (7.2) for the pure spin-up and spin-down cases, it follows *necessarily* from the linear character of the Schroedinger equation that the state emerging from the apparatus will be as given on the right side of the following measurement transition formula:

$$\Psi \to a(\uparrow \cdot M^+) + b(\downarrow \cdot M^-). \tag{7.4}$$

The interpretation of the state that emerges from the measurement is this: the probability that the spin is up and that the needle moves to the M^+ interval is $|a|^2$; the probability of spin down and needle in the M^- interval is $|b|^2$. Conspicuously, for a well-designed experimental setup obeying Eqs. (7.1) and (7.2), there are no terms $(\uparrow \cdot M^-)$ or $(\downarrow \cdot M^+)$. Concerning the two terms that do appear in Eq. (7.4), there is nothing that tells us which one materializes; that is, that tells whether the needle will settle in one interval or the other. There is no collapse of the wave function in the mathematics of Schroedinger's equation.

Of course, if you happen to know that the needle is in fact in a particular one of the intervals, say M^+, you will be able to make some safe bets concerning subsequent measurements on the neutron. You will bet (should bet!) that the neutron spin is up. You will proceed, that is, as if the wave function had indeed collapsed to the spin-up state. *But how did you know where the needle settled?* Does the chain of correlations become a measurement only when it finally reaches a sentient being who, acting as an outside observer, forces a selection? This possibility has been advocated by Eugene Wigner among others. It is a view that is hard to falsify but hard to build upon and hard to swallow without yielding to rank solipsism. Moreover, whose solipsism? Consider the case of *Wigner's friend*. Wigner wants to know which of the lights flashed, blue or red. He asks his friend, who has been observing. "The blue light flashed," the friend says. "Yes, but what was the result before I asked you?" Is it only at the stage when his friend's spoken answer enters Wigner's consciousness that the wave function has collapsed to spin up? Or did it collapse at the earlier moment when the blue flash registered in the consciousness of another sentient being, Wigner's friend?

Schroedinger's Cat

A celebrated variant was offered up whimsically by Schroedinger in a long article on the interpretation of quantum mechanics. Imagine a diabolical experiment in which a cat is penned up in a lidded cage along with a Geiger counter and a quantity of radioactive substance so tiny that the probability that one atom will decay in a period of one hour is just fifty percent. If an atom decays, the Geiger counter will respond, instantaneously setting off a chain of events that releases a quantity of hydrocyanic acid sufficient to instantly kill the cat. How does a quantum mechanician contemplate the situation when the hour is up and he is about the lift the lid? He has no choice but to attribute to the whole system—the cage and its contents—a wave function that has the cat in an equal superposition of states, dead and alive. That's weird! This superposition business may be all very well for atoms, but for a cat? Of course, an external observer who looks in after the hour is up will find one outcome or the other, dead or alive. But there is no known physical observation that corresponds to the superposition state; that is, the superposition state is not an eigenstate of any realistically imaginable observable. The observer (or a proxy such as a heart monitor) forces the selection between life and death in this grim tale. But what about the cat? How does she feel? Remember, she is not an outside observer. And was her fate really sealed only when the observer lifted the lid?

We are actually all of us, daily, in the position of Schroedinger's cat. When you cross the street against the light in busy traffic there is a non-negligible probability that you will be struck down and killed. To an outside observer who first checks up after the allotted time for crossing, you are in the cat's predicament; you are in a superposition of the states dead and alive. Much more generally, to an outside observer who checks up at some future time with respect to *any* kind of distinction (dead, alive; rich, middling, poor; bald, hirsute; etc.), we are all of us in states that are superpositions over the possible outcomes. That our fate is reckoned in probabilistic terms is in itself not surprising. That is familiar enough in everyday life. What is eerie is that to the outside observer we are superpositions until the observation is made.

Delayed Choice

Contemplate the setup shown in Fig. 7.1a. A beam of monochromatic light generated at the source S strikes a half-silvered mirror (beam splitter) at A. Part of the beam is reflected to an ordinary mirror at B; another part, say of equal intensity, is transmitted to the ordinary mirror at C. Photon detectors D_{I} and D_{II} are located as shown. The beam reflected from B goes on toward the detector D_{II}; that from C goes toward the detector D_{I}. If the source intensity is weak enough, the detectors will respond with distinguishable clicks, each click representing the arrival of a photon. Some detections are made by D_{II}, others

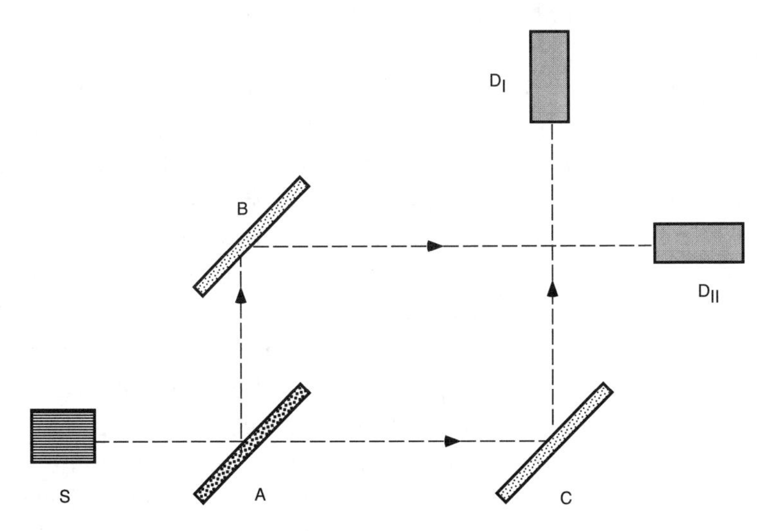

Figure 7.1a A setup, in connection with "delayed choice" issues, that brings out the particle-like aspect of phenomena.

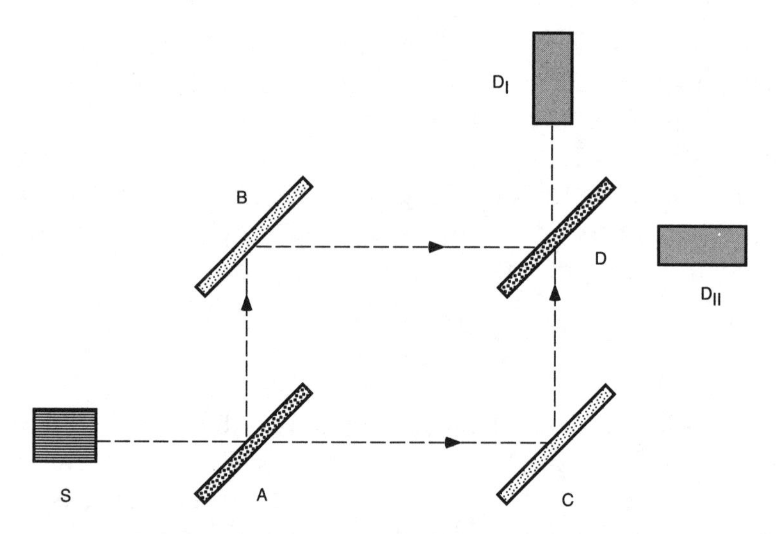

Figure 7.1b An alternate "delayed choice" setup.

by D_I. The odds are 50 : 50 as we've arranged it. For the former class of events, one will be inclined to say that the photon took the path $A \rightarrow B \rightarrow D_{II}$; for the latter class, $A \rightarrow C \rightarrow D_I$. This is a very particle-like picture of what is happening. The individuality of detections, whole clicks, tells us that light is composed of photons and that each photon takes one definite path or the other.

But now look at Fig. 7.1b. It's the same as Fig. 7.1a except that a half-silvered mirror D has been placed in the way of the beams, as shown. With this setup only one of the two detectors, namely, D_I, responds; D_{II} records no events at all! We have been through something very much like this earlier on, in connection with the double-slit experiment. What we are seeing with the half-silvered mirror inserted at D is the wavelike side of quantum mechanics. There is a probability *amplitude* for each of the two paths in the setup of Fig. 7.1a. If the half-silvered mirror at A does its job, the probabilities are 50 : 50. In the setup of Fig. 7.1b, the half-silvered mirror at D induces a shift in the relative phases of the two amplitudes so that they interfere destructively (as we have arranged it) at detector D_{II} and constructively at D_I.

The new twist for the class of experiments under discussion here has to do with *delayed choice*. Insert the (half-silvered) mirror D and listen as the counter D_{I} goes "click" from time to time. As for D_{II}, it never clicks. Now, on a sudden whim, remove the mirror D. The clicking of D_{II} will now be heard interspersed among clicks from D_{I}. But there is a quantum mechanical surprise in store. It can happen that D_{II}, but not D_{I}, will make a response very soon after the mirror has been removed, so soon that the photon about to be detected had already traveled most of the way from source to detector, therefore thinking it was in the setup of Fig. 7.1b. In this circumstance we might have expected that the photon had already committed itself, wavelike, to both paths. But in fact there is an early response only from D_{II}. The photon seems to have selected one path over the other. As John Wheeler has put it, in deciding when to remove the half-silvered mirror, we decide what the photon *shall have done* after it has *already done it*. Experiments of this sort have actually been carried out, although the description given above is merely notional. The half-silvered mirror is not in fact inserted or removed on a whim. Rather, a device that fulfills its role is always in place, either activated or not. Activation or deactivation is not done on a personal whim but is decided by a random number generator. When all is done, quantum mechanics emerges triumphant. We are again reminded that the features displayed by a quantum mechanical system depend on the experimental setup employed.

The EPR Argument

Einstein never gave up. In 1935, years after the main debates with Bohr had seemingly waned, Einstein, Podolsky, and Rosen (EPR) published a paper questioning whether the quantum mechanical conception of reality can be complete. The paper came as something of a bombshell for its time. The first sentence is worth quoting. "Any serious consideration of a physical theory must take into account the distinction between the objective reality, which is independent of any theory, and the physical concepts with which the theory operates." The authors went on to propose as a necessary requirement for a theory to be complete that "every element of the physical reality must have a counterpart in the physical theory." Then comes a key criterion concerning physical reality: "If, without in any way disturbing a system, we can predict with certainty (i.e., with probability equal to unity) the value of a physical quantity, then there exists an element of physical reality corresponding to this physical quantity."

If you are not on your quantum mechanical guard you may find that these dicta are highly plausible. But once alerted by EPR, you can easily check that they lead to results in conflict with quantum mechanics. This can be illustrated on a number of examples. EPR took the case of position and momentum measurements but it will be slightly easier here to focus on spin. Consider a system of two spin-one-half particles, for example, an electron and positron. Let $S_x(e)$, $S_y(e)$, $S_z(e)$ be the electron spin components along the x, y, z axes; similarly, let $S_x(p)$, $S_y(p)$, $S_z(p)$ be the corresponding spin components of the positron. Now, there is a particular state of the two-spin system, the so-called *spin-singlet* state, in which the total spin angular momentum is zero. It is a superposition of the two states shown below: one with electron spin up, positron spin down; the other with electron spin down, positron spin up. We symbolize this superposition state by

$$(\uparrow\downarrow - \downarrow\uparrow)/\sqrt{2}, \tag{7.5}$$

where the first arrow in each term refers to the electron, the second to the positron. Suppose that the particles are prepared in this spin state and allowed to fly far apart. At a certain instant you measure the component of the electron spin along some particular axis. At the same instant (your watches are coordinated) a confederate, far away, measures the positron spin component along the same axis. If you find up, your partner *must* find down; and vice versa. The odds are 50 : 50 as between these two possibilities.

But up and down with respect to what direction? The answer is, with respect to *any* space direction. If you and your partner are both measuring the component of spin along the z direction, then if you find up (down) along that axis, your partner must find down (up). So too if, instead, you are both measuring the x component of spin, or both measuring the y component, or the component along any direction whatsoever. How then does this fit with EPR? They would say that your measurement of the electron cannot have disturbed your partner's simultaneous measurement of the distant positron (since no signal from you—even if traveling at the speed of light—could have reached her in time to influence her measurement). Nevertheless, your finding leads to a definite prediction concerning your partner's measurement. She need hardly have bothered! If the electron spin is found to be up (down), the positron spin is necessarily down (up). According to EPR, the component of positron spin is therefore an element of physical reality; it can be predicted with certainty without disturbing the positron at all. But this is true for all three axes. Thus, whether for the positron or the electron (the above argument, of course, works in both directions; a measurement of the positron spin component leads to a definite prediction for the electron), S_x, S_y, and S_z are elements of physical reality. However, in the framework of quantum mechanics we know that the three components of spin do not commute among themselves. That is, there is no quantum state for which all three, or indeed any two, spin components can be known simultaneously. Therefore, according to EPR, there is something missing: quantum mechanics must be incomplete.

A great deal of ink was spilled in those early days over the EPR paradox, although now it does not seem much more strange than all the other oddities of quantum mechanics. One has to say simply (though it is rarely said simply) that the EPR notion of physical reality is too demanding for the quantum world that we actually inhabit. The main legacy of the EPR paper is that it injected the principle of *locality* into the analysis of measurements. This is the principle that a measurement here and now should not be able to influence a measurement elsewhere unless enough time has elapsed for a signal to get there, traveling no faster than the speed of light. We'll return shortly to this locality principle.

Hidden Variables, Bell's Inequality

From the early days of quantum mechanics, the question nagged: is there a deeper layer in which classical notions of reality prevail? This is the question of "hidden variables," the search for a dynamical foundation of quantum mechanics based on hidden, classical variables. On such a picture, at the deeper level any individual quantum system respects classical notions of reality. Here is John Bell's formulation: "To know the quantum mechanical state of a system implies, in general, only statistical restrictions on the results of measurements. It seems interesting to ask if this statistical element can be thought of as arising, as in classical statistical mechanics, because the states in question are averages over better defined states for which, individually, the results would be quite determined." Or, to quote Eugene Wigner: "The idea of hidden variables postulates that the description of states by the quantum mechanical state vector is incomplete, that there is a more detailed description, by means of variables now 'hidden,' which would be complete and the knowledge of which would permit one to foresee the actual outcomes of observations.... The relation of the postulated theory of hidden variables to present quantum mechanics would be similar to the relation of classical microscopic physics to macroscopic physics." An early analysis of the hidden variable issue was made by the great mathematician John von Neumann, who seemed to neatly dispose of the hidden variable program straightaway. He claimed to prove that hidden variables are necessarily incompatible with quantum mechanics. But this was subject to certain general conditions that he supposed could be placed on hidden variable theories. These conditions seemed reasonable enough at first; but they became suspect over time.

In the mid 1960s, John Bell revisited the hidden variable question with a more decisive insight and with a stunning result. It can be illustrated on the same two-spin system that we used above to describe the EPR paradox. Before turning to that, let us first consider the situation for a single spin-one-half particle, accepting that the underlying hidden variable dynamics can somehow account for the fact that the spin projection along any arbitrary direction can take on only the correct values $+1$ or -1 (this is in units of half the Planck constant). Which of these outcomes obtains in any particular case will depend on the particular values of the hidden variables. Indeed, the hidden variables are supposed to determine the outcome for the spin projections along all possible directions. In a hidden variable context, the spin projections in all possible directions are elements of physical reality. Despite this, to get around the EPR paradox we have to accept that the spin components in two (or more) different directions cannot be *known* simultaneously—the measurements disturb one another. But for a system of two spinning particles far apart, Bell assumes along with EPR that a measurement of a spin component of particle A cannot influence the outcome of a measurement of the same or any other spin component of particle B, provided the two measurements occur close enough in time so that a light signal could not have passed from one location to the other. As we have already said in connection with the EPR paradox, this *locality* hypothesis has the following consequence. For a system of two spins in the singlet state, a measurement on particle A of its spin projection along some particular direction automatically fixes the value of the spin along that same direction for the distant particle B. The spin projection of B is necessarily equal and opposite to that of A.

Bell's great idea was to consider spin projections not just along some one direction but, rather, along a set of directions. Three directions—call them a, b, c—will in fact do for our present purposes (these directions need not be orthogonal). Let us focus on the spin situation for particle B. With respect to the three directions, there are eight ranges of the hidden variables, corresponding to B spin projection up or down, denoted respectively by $+$ or $-$, for each of these directions. Let us denote these by the symbol (a, b, c), where each letter can take on the values $+$ or $-$. Thus, $(+, -, +)$ is the state where the spin projection is up along the directions a and c, down along b; and so on. The unknown probability distribution over the hidden variables translates to a distribution in the probabilities of the eight different spin possibilities (a, b, c). Denote the latter probabilities by $p(a, b, c)$. Thus, for example, $p(+, -, +)$ is the probability of the spin condition $(+, -, +)$; and so on.

Without any mutual interference, we can experimentally determine the spin projections of B along any two of these directions. We do this by performing one measurement directly on B, the other on the distant particle A. We can therefore find the probability [call it $P_{ab}(+, -)$] that the spin of particle B is up along a, down along b; and similarly for other two-direction probabilities such as $P_{bc}(+, -)$, $P_{ac}(+, -)$, $P_{ab}(+, +)$, and so on. It will be enough here to focus on the probabilities $P_{ij}(+, -)$ for the three pairs $(i, j) = (a, b), (b, c), (a, c)$. Clearly,

$$P_{ab}(+, -) = p(+, -, +) + p(+, -, -),$$

$$P_{bc}(+, -) = p(+, +, -) + p(-, +, -),$$

$$P_{ac}(+, -) = p(+, +, -) + p(+, -, -),$$

From these equations we deduce that

$$P_{ab}(+, -) + P_{bc}(+, -) = P_{ac}(+, -) + p(+, -, +) + p(-, +, -).$$

Since the probabilities $p(a, b, c)$ are inherently non-negative, it follows that

$$P_{ab}(+, -) + P_{bc}(+, -) \geq P_{ac}(+, -). \tag{7.6}$$

This is Bell's inequality as applied to the two-spin system. It should be obvious that the inequality states that the sum of *any* two of the three probabilities is greater than or equal to the third. It is purely a matter of notation that we have singled out $P_{ac}(+, -)$ to be placed on the right side of the above equation.

What we have actually presented here is Wigner's variation on Bell's theorem. Bell deals with averages, Wigner with probabilities. We will nevertheless refer to Eq. (7.6) as Bell's theorem.

His was the great breakthrough. What has gone into Bell's theorem is nothing much more than the principle of locality described above. It seems hard to quarrel with this assumption in a context of hidden, classical variables.

The probability $P_{ij}(+,-)$ clearly depends only on the angle θ_{ij} between the direction vectors i and j, so we may write $P_{ij}(+,-) = P(\theta_{ij})$. Thus, Eq. (7.6) may be written

$$P(\theta_{ab}) + P(\theta_{bc}) \geq P(\theta_{ac}). \tag{7.7}$$

Is this prediction compatible with quantum mechanics? The answer is that it is not! Quantum mechanics produces a definite formula for the probability function $P(\theta)$. Unfortunately, although the quantum calculation is straightforward, it requires somewhat more technology than we have developed. We therefore simply quote the result:

$$P(\theta) = \frac{1}{2}\sin^2\left(\frac{\theta}{2}\right). \tag{7.8}$$

It is now easy enough to check that for a wide range of choices for the three direction vectors, Bell's inequality, Eq. (7.7), is violated if $P(\theta)$ obeys the quantum formula of Eq. (7.8)! The conclusion: local hidden variable theories cannot provide a foundation for quantum mechanics. On the experimental front, tests of Bell's inequality have been carried out not just with material particles (protons) but also with photons, whose states of polarization are like the states of spin. The experiments are difficult and have had a history of ups and downs, but by now quantum mechanics has safely emerged the winner.

As said, hidden variables, in addition to all the other hurdles they face, are shown to be incompatible with quantum mechanics—unless one is prepared to relax the general conditions, notably locality, that go into the Bell theorem. In the 1950s David Bohm in fact succeeded in constructing an internally consistent hidden variable theory for a nonrelativistic particle; but it is highly nonlocal and, in any case, rather forced.

Given the successes and internal consistency of quantum mechanics, it seems a forlorn hope that one can ever return to classical notions of reality. If enlargements and revisions are going to be encountered in the future it is likely that these will carry us farther from, rather than closer to, our everyday intuitions. This could conceivably happen at the frontiers where quantum ideas seek to mesh with general relativity; or perhaps, some would say, where the quantum meets with consciousness.

Roundup

The formalism of quantum mechanics established itself early on. So too did the workaday rules for correlating the mathematical scribblings with empirical observations. On the mathematical side, the general framework seems to be thoroughly self-consistent. Empirically, quantum mechanics is hugely successful; there are no known contradictions. So, what more can one ask? Well, it would be nice to receive some aid and comfort in coping with the various oddities that quantum mechanics entails, of the sorts that have been presented in this and earlier chapters. Above all, we would like to understand how it is that probabilities become facts.

The hidden variable view is that quantum mechanics is incomplete, that classical reality prevails at a deeper level, presently inaccessible. New physics would appear if we could get at those variables observationally. That would be exciting indeed. But the hidden variable idea runs up against the Bell inequality. An alternative viewpoint in a direction opposite to hidden variables is, to quote Wigner, "that the function of quantum mechanics is not to describe some 'reality,' whatever that term means, but only to form statistical correlations between subsequent observations." This is not to deny, Wigner says, that there is a world out there outside of ourselves (whatever that means!). That world is awash in facts that have already established themselves. Quantum mechanics tells us which facts are possible (the eigenvalues) and which are not.

But within quantum mechanics itself, there seems to be an unbridgeable divide between the future and the present instant (and past instants insofar as we can reconstruct them from the record). The future is intrinsically statistical, with probabilities governed by the equations of quantum mechanics. The trouble is that this way of looking at the situation seems something of a cop-out. In effect, it abandons the idea of *explaining* how facts come about, taking as the main function of science merely to correlate them. When a fact in fact happens, the quantum mechanical wave function is simply declared to have collapsed; after all, it is only a correlational tool! And that's that. The more orthodox *Copenhagen interpretation* places the emergence of a fact at the point where it is first registered by a "classical" measuring instrument; that is, by a "large" apparatus in good working order. As a practical matter, this is in some sense undoubtedly the case. The meter readings are facts. But it has never been clear within the Copenhagen view how the meter makes its selection when there are multiple choices to be made. We may also recall here the notion, mentioned earlier, that facts first emerge only when registered in the consciousness of sentient beings, the ultimate measuring instruments! But having recalled it, there seems nothing much more to say.

Finally, we may briefly mention the so-called *many-worlds* interpretation of quantum mechanics. Proposed in 1957 by Hugh Everett III, it confronts the selection conundrum in a most audacious manner. Whenever there is a choice to be made among alternative outcomes of a measurement, the world splits into many worlds, all the possible outcomes emerging, one in each of the newly created worlds! This has been going on for a long time, of course, so there is a vast proliferation of worlds out there living side by side. But they are totally out of contact with one another. It's hard to know what to make of such an interpretation of quantum mechanics. As with the consciousness hypothesis, it cannot be falsified or built on. But it is undoubtedly amusing to contemplate. Each of us has clones all over the place, but we never meet.

The literature on the interpretation of quantum mechanics is vast and growing. The capsule comments of this brief chapter have scarcely summarized, let alone adequately expounded, all the main lines of discussion and investigation. Everybody has weighed in: philosophers, physical scientists, science journalists, talk show hosts, theologians,... (the list is alphabetical!).

At the end of the day, quantum mechanics remains both intact and puzzling.

CHAPTER EIGHT

The Building Blocks

We have mainly focused so far on the application of quantum principles to systems of immutable, nonrelativistic particles. In that framework, the various kinds of particles that occur in nature, as well as the force laws that describe their interactions, have to be accepted as inputs. In the case of electromagnetism and gravity, those force laws of course have a classical heritage. Nevertheless, in the nonrelativistic quantum context they entered from the outside. There is no inconsistency in any of this, but there are problems and limitations when one seeks to extend the framework. For one thing, it is not possible to achieve a self-consistent, relativistic generalization along the lines we have been following. Dirac's relativistic equation for the electron is enormously successful, but already for a single-particle system it harbors indications of its own conceptual limitations. Moreover, there is no provision in our treatment so far for particle creation and destruction, relativistic or otherwise.

A potential resolution of these difficulties presented itself early on after the birth of the "new" quantum mechanics. This had to do with the application of quantum principles to *fields*, the first such application being to the electromagnetic field system. Classically, particles and fields stand on an equal footing as dynamical systems. But when the electromagnetic field was first subjected in the late 1920s to a quantum treatment, something remarkable emerged. The quantized electromagnetic field yielded up those massless quanta, the photons, that were first intuited by Einstein in 1905. The photons were not put *into* the theory as particles, they came *out* on their own. The discovery that particles can emerge from fields led in time to a broad generalization: the notion that electrons and protons and, later, the various other particles we will be discussing, might also be understood as the quanta of corresponding fields. For all but the photon, the fields in question are unknown to us in any classical guise. They are invented as quantum fields de novo for the very purpose of yielding the wanted particle-like quanta. From the perspective of quantum field theory it is these fields, not their quanta, that are the fundamental theoretical entities. The questions, what are the fundamental particles of the world and what are the forces among them? are hereby replaced by the questions, what are the fundamental fields of the world and how do the fields interact among themselves? The notion of field *interactions*, as it translates into interactions among particles, is something we will turn to later on. But first we will visit the particle building blocks themselves, keeping in mind that the fundamental particles of one era can come to be regarded as composites in a later era.

There may have been a moment in the early 1930s when it could seem that all the basic building blocks of the whole wide world were at last in hand. The electron had been discovered in the final years of the nineteenth century; the proton achieved its identity as the nucleus of the hydrogen atom when Rutherford established his model of the atom a little over a decade later; the neutron was discovered in 1932, though it took a little while before it was accepted as a distinct new particle rather than a bound state of proton and electron. The photon had a protracted birth beginning with Einstein in 1905, followed much later by its reappearance, full blooded, as the quantum of the quantized electromagnetic field. So, there it was: material things are made of atoms, atoms are made of electrons and nuclei, nuclei are made of protons and neutrons; and there's light, made of photons. The whole world is reduced to electrons, protons, neutrons, and photons! Not bad. This triumph of reductionism was short lived, however. At just about the time that the neutron was discovered, a tiny bit earlier in fact, the positron turned up. This antiparticle to the electron had first arisen theoretically as an unanticipated consequence of Dirac's relativistic quantum equation of the electron. Once the positron materialized it seemed probable to many that the proton and neutron ought to have their own antiparticles as well. They do. The antiproton and antineutron were discovered in the 1950s. Also in the early 1930s, indeed, even a bit before the positron's discovery, the neutrino was postulated. According to Pauli it was needed to save the principle of energy conservation in nuclear beta decay. In that reaction a parent nucleus decays to a daughter nucleus, ejecting an electron in the process. But the electron carries off only a (variable) fraction of the energy available. Pauli's proposal was that the missing energy is carried away by an unseen neutral particle. The kinematic data on beta decay indicated that this particle must have a very tiny mass, if any mass at all. Adopting the general ideas of quantum field-theory that were proving so useful for the quantum theory of electromagnetism, in 1933 Fermi devised a field theoretical account of beta decay starring the neutrino. This was all very momentous for the times. One did not then lightly postulate new particles or introduce new quantum fields. The theory correctly predicted that neutrinos interact with matter only very feebly. Their direct detection had to await heroic experiments first carried out in the mid 1950s at a copious neutrino source, the powerful Savannah River nuclear reactor in Georgia.

In the midst of the other discoveries and developments of the early 1930s, attention began to focus increasingly on the nature of the forces that hold the proton and neutron ingredients of nuclei together. The Coulomb force won't do. It does not act on neutrons, which are neutral; and it is repulsive between pairs of protons. Moreover, it was clear that nuclear forces must be considerably stronger than the Coulomb force, although of very short range: the former because the nuclear ingredients are typically much more tightly bound than are electrons in an atom; the latter because those nuclear ingredients are held together in a volume that is tiny on the scale of an atom. In 1934, the Japanese physicist Hideki Yukawa entered new territory by proposing a field-theoretic account of nuclear forces. The theory led him to predict the existence of two new particles: positive and negative pions, π^+ and π^- as we now denote them. The one is antiparticle to the other and they have identical masses. The theory entails a correlation between the mass and range of the nuclear force. In Yukawa's rough treatment, the force between a proton and a neutron corresponds to the potential

$$V(r) = -g^2 \frac{e^{-r/R}}{r},$$

where g is a "coupling" constant and R a "range" parameter. Owing to the exponential function, the potential begins to fall off rapidly for $r \gg R$. In this sense the potential is said to be of range R. In Yukawa's theory, the mass m_π of the pions was related to the range parameter by

$$R = \frac{\hbar}{m_\pi c},$$

where c is the speed of light. From nuclear information concerning the range, Yukawa was able to make a crude estimate of the mass of the pions: $m_\pi \approx 200 m_e$, where m_e is the mass of the electron.

Within a few short years, evidence for new charged particles with masses intermediate between those of the electron and the proton began to show up in cosmic ray experiments, and they soon came to be identified as possible candidates for Yukawa's pions. Ten years of confusion then ensued. The new particles were unstable, which was OK; and the mass, though not initially well pinned down, accorded well enough with Yukawa's rough prediction. But other properties did not make sense. The decay lifetime was much longer than expected; and, most tellingly, the absorption characteristics as these particles pass through matter did not fit expectation: the putative pions interacted with nuclei far too feebly. In 1947 a way out was proposed: Yukawa's mesons exist but they decay into another species which is longer lived and weakly interacting; and it is the latter particle that is predominant in the cosmic rays at the lower altitudes where the cosmic ray sightings were first made. According to this proposal, it was the daughter particle that had been detected. Sure enough, at just about this time the experimental situation began to sort itself out in high-altitude cosmic ray experiments employing photographic emulsions to record the tracks of charged particles. There are indeed two distinct species of charged particles involved: Yukawa's pions $\pi^{\pm}$ and, somewhat lighter in mass, particles that we now call muons, $\mu^{\pm}$ (the μ^{+} being antiparticle to μ^{-}). Confirming evidence soon emerged at the large new particle accelerators that were coming into existence after the war. As we now know, π decays to μ plus a neutrino. The muons in turn decay according to the scheme $\mu^{\pm} \rightarrow e^{\pm}$+neutrino+antineutrino. The symbol e^{+} stands for the positron, e^{-} for the electron. The pion mean lifetime is 2.6×10^{-8} sec. Its rest mass energy is $m_{\pi}c^{2} = 140$ MeV. The muon lifetime is 2.2×10^{-6} sec. Its rest mass energy is $m_{\mu}c^{2} = 106$ MeV. For reference, we may note that the electron rest mass energy is $m_{e}c^{2} = 0.511$ MeV, the proton's 938 MeV. In later discussions we will replace the proper term "rest mass energy" by the shorthand word "mass," so masses will be given in energy units. Also note that in speaking of the lifetime of a particle, we are referring to the *mean* lifetime as measured in the particle's rest frame. The masses and lifetimes quoted above are recorded here only to a few significant figures. They are now known to a much higher level of accuracy. Protons and neutrons, the ingredients of atomic nuclei, are collectively called *nucleons*. Interest in the forces among nucleons (proton-proton, neutron-neutron, proton-neutron) intensified rapidly in the 1930s following the discovery of the neutron and the introduction of Yukawa's meson theory. It was not long before the meson hypothesis was extended to the prediction of a neutral counterpart to $\pi^{\pm}$, the so-called π^{0} meson or neutral pion. It was discovered in 1950. The mass is very close to that of the charged pions, as was predicted. It decays into two photons with a mean lifetime of about 10^{-16} sec.

Before turning to the onslaught of still other discoveries that got underway in the early postwar years, let us pause to review the humble collection of building blocks we have already assembled: electron, proton, neutron, and their antiparticles; photon; neutrino and antineutrino; charged and neutral pions; muons (particle and antiparticle). For most of science and technology, the *effective* building blocks are the electron, photon, and a large collection of different atomic nuclei, many hundreds of them. For most purposes the nuclei can be treated as tiny, almost pointlike objects characterized sufficiently well by charge Ze, magnetic moment, and mass. The atomic charge number Z is the most important of these parameters; it is what distinguishes one chemical element from another. Many of the elements come in a variety of isotopes with nuclei sharing the same atomic charge Z but differing in mass. With the discovery of the neutron it became clear that nuclei are made of protons and neutrons, the atomic charge Z being the number of protons and the nuclear mass very nearly proportional to the total number of nucleons (protons plus neutrons). It was a great conceptual advance that the vast array of known nuclei could be reduced to combinations of just two building blocks, protons and neutrons. The "everyday" world is thereby reduced to electrons, protons, neutrons, and photons.

But what about the other objects on our list? The antielectron (positron) and by extension the antiproton and antineutron got onto the list before they were discovered experimentally. They emerged, initially unanticipated, out of Dirac's attempt to find a proper relativistic quantum equation for the electron. The neutrinos and antineutrinos were hypothesized on the basis of somewhat more phenomenological considerations, as entities that carry off the energy that seemed to be lost in beta decay processes. Beta decay constituted a first beachhead outside of quantum electrodynamics for the newfangled ideas of quantum field theory. As we now know, there are in fact three different species of neutrinos and corresponding antineutrinos. The pions on our list were invented in a field theory context in a first attempt to account for the forces between nucleons, forces that govern the properties of atomic nuclei. Of all the particles on our list, it was only the muons that showed up without advance notice or obvious "usefulness." As we now know, the muon is in some sense much like the electron, with the decisive exceptions that it is about 200 times heavier and that it is unstable, with a lifetime in its rest frame of only a couple of microseconds. Here, and sometimes elsewhere, we use the term "muon" in a collective sense to include both μ^+ and μ^-; so too the terms "electron," "neutrino," "proton," and so on, are often used collectively to include both particle and antiparticle.

As said, the muons and charged pions were discovered, entangled at first, in cosmic ray experiments. Earth is constantly bombarded by energetic particles coming from outer space, with energies ranging upward to at least 10^{20} electron volts! There is a substantial flux of modest-energy (order of an MeV) neutrinos and protons whose origin is the sun. The neutrinos do not interact much with the atmosphere, or with the whole solid earth for that matter; they mostly pass on through. The more energetic cosmic ray particles come from farther out in the universe. Cosmic ray interactions in the atmosphere are predominantly initiated by the incoming proton component. The incident protons collide with the nitrogen, oxygen, and other nuclei in the atmosphere, knocking neutrons and protons out of the nuclei and producing pions and other particles. The ejected nucleons, pions, and other products of the initial collisions generate secondary collisions, although sometimes the unstable secondaries decay into other particles. For example, charged pions sometimes decay into muons and neutrinos before they have had a chance to undergo secondary collisions. Neutral pions generally do not live long enough to collide at all. Once produced in collisions, they decay quickly to photons. The latter collide with atmospheric nuclei to dislodge nucleons and produce electron-positron pairs, pions, and other particles. Sometimes, the positrons annihilate against electrons in the atmosphere to produce photons. And so on—primary collisions, secondary collisions, tertiary collisions, decay processes. Altogether, the atmosphere is the scene of complicated cascades of events that generate fluxes of all the various particles that appear on our list, and more to come!

The cosmic ray atmosphere has served splendidly as a high-energy physics laboratory for many, many decades, although for most (but not all) issues in particle physics it has long since been superseded by man-made particle accelerators. That transition began to take place in the early 1950s, but not before other major discoveries had been made in the cosmic ray venue. In 1947, two events recorded in a cloud chamber exposed to the cosmic radiation provided evidence for two new particles: one, a neutral particle of mass about 500 MeV that decays to a pair of charged pions, π^+ and π^-; the other, a charged particle of about the same mass that decays into one charged and one neutral pion. This development was a case of pure discovery, unforeseen, and it took a brief while for the findings to be absorbed. Then, quickly, the pace intensified. More and more new particle species began to show up; for the first few years exclusively in cosmic ray experiments typically employing cloud chambers or photographic emulsions, then increasingly at the new high-energy accelerators that were coming into operation. The early trickle turned into a flood.

Almost three hundred particle species have been identified by now! Almost all are unstable against spontaneous decay. Indeed, to the best of current knowledge the only stable species are the electron and proton and their antiparticles, the photon, and the neutrinos and their antiparticles. If not destroyed in collisions, all the others eventually decay into sets of the stable species, either directly or through unstable intermediaries. "Eventually" can be a very short time indeed, as little as 10^{-24} seconds for some species. Even the neutron is unstable in isolation, although it is energetically stabilized against decay when bound in a stable nucleus.

This onslaught of new discoveries opened up a new subnuclear world. The components of our everyday life—photons, electrons, protons, and neutrons—are somehow joined in a wider framework to a host of fellow particles, most of them transitory. The great challenge was, and remains, to search for patterns in their properties and mutual interactions and thereby to discover the fundamental laws governing their exis-

tence and behavior. For our purposes, the story of the building blocks can be divided into several overlapping epochs. The first extends from antiquity through the growth of the atomic hypothesis and discovery of the electron, and on to the late 1940s. It produced the ingredients of the atom and its nucleus as well as the other objects on our earlier list; some of them, although not actually discovered until later, at least postulated on the basis of strong experimental or theoretical evidence. The second epoch dates from the beginning of the new particle flood described above. It ushered in an age of rampant discovery, not just of new particles but of various regularities and patterns that began to show up in the data. At a deeper level there was impressive theoretical success in certain limited areas, notably quantum electrodynamics; and there were important insights gained on many other fronts as well. By the late 1960s, inaugurating a third epoch, the various strands of understanding and speculation began to draw together into the comprehensive quantum field theory that rules today, the so-called *standard model*. The impulse for its construction came from various quarters; above all, from some highly speculative ideas introduced many years earlier concerning a special class of quantum field theories, so-called *gauge* theories. Vital too, especially for the present story line, was the introduction of the quark hypothesis in the early 1960s. On the experimental side, a crucial initiating role was played by a set of experiments carried out in the late 1960s on the scattering of very high-energy electrons off protons and neutrons. Over the subsequent years the theoretical picture was refined and consolidated, guided in part by a string of further, often dramatic, experimental confirmations and discoveries. The standard model is now solidly established, although for all its success it is not yet at the end of the road. We are presently embedded in a fourth epoch, seeking to go still deeper.

Particles in Collision, Particles in Decay

Nature reveals itself not just through the particles that *are* but through the things that they *do*. There are two broad classes of things that particles do. (i) The unstable ones—and they are in the majority—decay spontaneously, transforming into sets of other, daughter particles. Especially for the heavier unstable parents, there can be many competing decay reactions. (ii) When two or more particles are brought into collision they can scatter, unchanged in identity and unaccompanied by any additional particles. But depending on the energy they can also transform into different sets of particles. At high energies there is generally a host of such competing reactions.

We will start with this broad class of collision phenomena. To begin with a specific example, consider what happens when two protons collide. If the energy is very small, the dominant reaction is "elastic" scattering, $p+p \rightarrow p+p$, the same particles in and out. At higher energies there are competing processes in which one or more pions are produced in company with the outgoing pair of nucleons, in some cases a proton having been transformed into a neutron. At still higher energies more and more reaction *channels* enter into competition (the set of particles produced in any particular reaction constitutes a channel); and so too for processes initiated when other pairs of particles collide, for example, electrons and positrons, pions and protons, and so on. At the highest accelerator energies so far achieved, protons and antiprotons with energies of close to one trillion electron volts each are brought into head-on collisions. At these extreme energies many hundreds of competing channels are open, some with particle multiplicities reaching into the hundreds!

Collision Cross Sections

Quantitatively, collision reactions are characterized in terms of the concept of *cross section*. To illustrate, take the case of a projectile particle incident on a target particle initially at rest, say a pion incident on a target proton. At any energy, however small, elastic scattering is always a possible reaction. But especially at higher energies, other more complicated channels compete. The probability of any particular reaction can be expressed in terms of a cross section, defined as follows. One imagines a disc that the target presents to the incident particle. The target is to be thought of as a point sitting at the center of the disc, the projectile as a point particle approaching on a straight line perpendicular to the surface of the disc. If that line crosses the disc, the reaction is said to occur. If not, not. The disc can equivalently be associated with the projectile: if the moving disc encompasses the target particle, the reaction occurs. In either way of looking at things, the area of the disc defines the cross section for the particular reaction in question. If you have a known flux of projectiles incident on a known density of target particles, knowledge of the cross section permits you to compute the rate at which reaction events of the type in question will occur. Conversely, one deduces cross sections experimentally by measuring reaction rates. Each competing reaction has its own characteristic cross section, the various cross sections depending in general on the energy of the collision. For a given pair of particles in collision, the sum of all the competing cross sections constitutes the *total* cross section. The latter determines the net rate of events of any kind.

These disclike concepts are not to be taken literally as corresponding to actual physical, intercepting objects associated with either the target or projectile particle. Rather, cross sections are a picturesque way of quantitatively characterizing the propensity for various reaction processes to occur. The bigger the cross section, the bigger the propensity. For proton-proton collisions at 100 GeV incident energy, the total cross section in round numbers is $\sigma_{\text{total}} = 4 \times 10^{-26}$ cm^2. This corresponds to a hypothetical disc of radius about 10^{-13} cm. It turns out that this total cross section value is roughly typical in that energy region for a wide class of colliding partners, including pion-nucleon, nucleon-nucleon, nucleon-antinucleon, and others (recall that protons and neutrons are, collectively, nucleons). There is another class of collision processes that at comparable energies have noticeably smaller cross section; for example, the collision of electrons and protons. And there are yet other cross sections that are far smaller still. We will return later on to these patterns of collision propensity or *strength*.

The array of distinct, imaginable collision reactions is enormous. Given N different particle species there are $N(N+1)/2$ possible pairs of collision partners, a very large number considering that N is not far from 300. Moreover, for any one pair of colliding particles there can be many competing reaction channels, a number that increases unlimitedly (so far as we know) with increasing collision energy. Thus, for our earlier example of proton-proton collisions, at very low energies the only important process is elastic scattering, $p + p \to p + p$. At somewhat higher energies it becomes energetically possible to create a single pion: $p + p \to: p + p + \pi^0$, $p + n + \pi^+$. At still higher energies, two pions can be produced: $p + p \to: p + p + \pi^0 + \pi^0$, $p + p + \pi^+ + \pi^-$, $p + n + \pi^0 + \pi^+$, $n + n + \pi^+ + \pi^+$. At higher and higher energies, more and more particles (and not just pions) can be, and in fact are, produced in reactions that compete with these more parsimonious ones. At the highest currently available accelerator energies, there are reaction channels containing hundreds of particles, mixtures of nucleons, pions, K mesons, and others. To add to the cornucopia, consider also that any particular reaction is characterized not just by its cross section but also by its dependence on collision energy and by the angular and energy distributions of the produced particles.

In the face of this vast richness of phenomena, the great trick is to search for general patterns and regularities; and to focus on those special features in the data that are likely to be most diagnostic and informative about the underlying science. There has been great progress along these lines, as we will discuss.

Lifetimes, Branching Ratios

The other thing that particles do, all but the stable ones, is decay. As has already been said, the decay process is essentially exponential in time, the net decay propensity being characterized by an average lifetime (or, simply, "lifetime"). It is the analog of the total cross section for collision reactions. The smaller the lifetime, the greater the decay propensity. Where there are competing modes of disintegration, the individual channels can be characterized by their *branching ratios*. The branching ratio for any particular decay mode is the fraction of all decay events that proceed via that particular channel.

The number of competing decay channels that are available is limited in part by energy conservation. Since it takes energy to create mass ($E = mc^2$), the heavier unstable particles, with their larger rest energies, typically have more channels open to them than do the lighter ones. For example, the charged D meson (mass = 1870 MeV) has dozens of major decay modes and many minor ones as well. Charged pions (mass 140 MeV) have only a single major decay channel, pion → muon + neutrino. It must be emphasized that minor decay modes are by no means without interest. For example, the charged pion decays not only in the manner noted above, but also into an electron and a neutrino with a minuscule branching ratio of about 10^{-4}. Discovery of this rare process played an important role in the development of our understanding of the so-called weak interactions. Rare collision and decay reactions are in fact often at the center of attention. Their very rarity presents a daunting experimental challenge, however. Modern techniques allow one to press searches for rare events down to branching ratios of order 10^{-10} in certain favorable cases.

Accelerators

Modern, high-energy particle accelerators descend from several lines of development pioneered in the late 1920s and early 1930s; most notably the cyclotron. Present-day machines dwarf their ancestors in size, complexity, and energy; but the basic plan is always the same: electric fields are employed to accelerate charged particles to high energies. In *linear* accelerators this is accomplished in one pass through the device. In *circular* machines (Lawrence's great idea), a magnetic field constrains the particles to move around and around in a circular orbit, allowing for multiple passes through an electric field. Typically, accelerator complexes nowadays employ combinations of both kinds of devices, which can be used independently and in sequence as particles are accelerated from low to very high energies.

In so-called *fixed target* setups, the high-energy particle beam produced by the accelerator is caused to impinge on a condensed target, solid or liquid. For the very high-energy processes of concern here, the forces that bind the ingredients of the target atoms together are negligible, so for many purposes the target can be regarded as a bag of independent protons, neutrons, and electrons. If a given collision event is sufficiently well measured, it is generally possible to determine whether the target particle was a proton, neutron, or electron. In the class of machines known as *colliders*, instead of a single beam incident on a stationary target, two separate beams are accelerated to high energy and caused to come into essentially head-on collision. One or both of the two beams can also be used individually for fixed target experiments.

The two kinds of setups, fixed target and collider, have their separate merits. For definiteness, consider the collision of two particles of identical mass m, say a proton-antiproton or electron-positron collision. Let E be the total laboratory energy, kinetic plus rest energy, of a beam particle. In the typical collider setup, two particles moving with equal but opposite momenta collide head on. The net momentum is therefore zero and the net energy is

$$W_{\mathrm{C}} = 2E.$$

The subscript C symbolizes that we are dealing with a collider setup. This energy is shared among the reaction products, some of it being incorporated in their rest energies, what's left over going into the kinetic energy of motion of the products. The net momentum, summed vectorially over all the reaction products, remains zero. In the fixed target setup, a beam particle of energy E collides with a target particle at rest, so the net energy in the laboratory frame is $E + mc^2$. For purposes of comparison with the collider situation, it is convenient to ask what energy would be seen in the *center of mass* reference frame of the collision. This is a frame moving in the direction of the beam particle at a speed such that an observer in that frame sees the colliding particles as having equal and opposite momenta. In the center of mass frame, the collision looks just like a collider event. It is easy to work out the net energy in this frame. It is

$$W_{\mathrm{FT}} = \sqrt{2mc^2(E + mc^2)}.$$

The subscript symbolizes that this is the center of mass energy corresponding to a fixed target collision event in which the projectile has energy E in the laboratory frame. The main thing to notice is that W_{FT} is smaller than W_{C} at all energies E. It is in fact very much smaller if $E \gg mc^2$. Equivalently, the energy W_{FT} in the center of mass frame is smaller than the energy $E + mc^2$ in the laboratory frame. In contrast, for a collider setup the laboratory and center of mass frames are one and the same. The important thing is that it is only the center of mass energy that is fully available for creating rest mass.

Why this difference between fixed target and head-on collisions? The answer lies in energy-momentum conservation. In the fixed target setup the incident projectile not only supplies kinetic energy, it also brings in momentum. But momentum must be conserved in the collision. The reaction products therefore have to carry it off and hence carry off motional (kinetic) energy. That motional energy is "wasted" in the sense that it is not utilized to supply the rest energy needed to create particles. In contrast, in head-on collisions the *net* momentum is zero. The full energy W_{C} is therefore available for incorporation in rest mass and hence for creating particles. To illustrate, consider the reaction

$$p + p \to p + p + X,$$

where X is a particle of mass M. Take the mass of the proton to be m. How much energy has to be supplied to the colliding particles to reach the energy threshold for this reaction? In the collider setup each proton of the colliding pair is invested with *kinetic* energy that we shall denote by K_C. Clearly, the threshold kinetic energy is $K_C = Mc^2/2$. At that incident energy, the reaction products emerge at rest. For the fixed target setup, let K_{FT} be the incident proton's kinetic energy at threshold. It is easy to check that the ratio of beam particle kinetic energies for the two setups is

$$K_{FT}/K_C = M/m + 4.$$

This is a ratio never smaller than 4. It is much larger than that if $M \gg m$. Thus, if the rest energy of the X particle is 100 times that of the proton, the collider threshold is about 50 GeV, the fixed target threshold about 5000 GeV!

Colliders therefore have the greater reach for discovery of high-mass particles. But fixed target machines have their own advantages. Once one is above the threshold for any particular reaction, whether in a fixed target or collider machine, there will be a spectrum of kinetic energies of the produced particles. For given beam energy, that spectrum in general reaches to higher values in the fixed target case. Insofar as these reaction products are used to induce secondary collisions, the higher their energies the better. Fixed target setups have another advantage. A beam of projectile particles sees a much greater density of available collision partners in a condensed target than it does in another beam approaching it head on. The particle density in beams, that is, is very much smaller than in solids or liquids. Total event rates at fixed target machines are therefore generally much higher than at colliders. For example, the 30 GeV proton beam at Brookhaven's Alternating Gradient Synchrotron (AGS) accelerator generates several *trillion* collision events per second on a solid target. The colliding 900 GeV proton-antiproton beams at Fermilab's Tevatron facility produce somewhat less than a *million* events per second.

What kinds of charged particles are available for acceleration at high-energy facilities? Of the hundreds of known species, the only ones that are readily at hand to start with are the electrons and protons that go to make up atoms; or, for some purposes, various atomic nuclei regarded as coherent entities.

Altogether, therefore, the kinds of beams that can be contemplated for the first stages of any accelerator complex are beams formed of electrons, protons, and various atomic nuclei. The ingredients of condensed targets, electrons, protons, neutrons and atomic nuclei can also serve as collision partners in a fixed target setup. Altogether, this allows for a considerable variety of partnerings: electron-electron, electron-proton, proton-proton, neutron-proton, nucleus-nucleus, and so on. All have been pursued. Beyond this, particles of other types created in primary high-energy collisions can themselves, if they live long enough, be brought into secondary collisions; and so too for their decay products. In this way, secondary beams of photons, neutrinos, positrons, antiprotons, pions, *K* mesons, muons, and other species of charged and neutral particles have become available for fixed target experiments. For example, neutrino-proton scattering experiments get their neutrinos largely from pion decays, the pions themselves having been produced at fixed targets bombarded with energetic proton beams. Some of these secondary particles can also be used to form one of the two beams in a collider. For that application the secondaries have to be collected, stored, and boosted in energy. This requires that they be long lived and charged, requirements that presently limit such applications to positrons and antiprotons as additions to the list of beams available for colliders. When you read about proton-antiproton colliders, for example, you should know that those antiprotons are collected from the debris produced when a proton beam impinges on a condensed target. So too, electron-positron colliders get their positrons from the debris produced by an electron beam incident on a solid target.

There are nine major accelerator centers in the world today: Fermilab, Stanford, Cornell, and Brookhaven in the United States; CERN (Geneva) and DESY (Hamburg) in western Europe; KEK in Tsukuba, Japan; the Institute of High Energy Physics in Beijing, China; and the Budker Institute in Novosibirsk, Russia.

Fermilab's Tevatron is the world's highest-energy particle accelerator. It accelerates protons and antiprotons to 900 GeV and operates in both collider and fixed target modes. In the former, the center of mass energy is of course $2 \times 900 = 1800$ GeV. In the latter, the center of mass energy is much less, about 40 GeV; but the fixed target operation generates valuable secondary beams of neutrinos, pions, muons, and other species. The world's most intense high-energy proton beam is located at Brookhaven's AGS, a 30 GeV proton fixed target facility. A heavy ion collider, RHIC, will soon come into operation at Brookhaven.

The highest-energy electron-positron collider is the CERN accelerator LEP, a circular machine with a 26 kilometer circumference. Each beam has an energy of about 90 GeV. In the middle of the first decade of the twenty-first century, a proton-proton collider will come into operation inside that ring with beam energies of 7 TeV. That's about seven times the Tevatron energy! CERN also operates a fixed target facility employing 440 GeV protons.

The SLC facility at SLAC (Stanford) is an electron-positron collider with beam energies of 45 GeV. It has the distinction of being the world's first and only high-energy *linear* collider (all the others are circular devices), and may be a harbinger of bigger linear machines to come. Other electron-positron colliders operate in Japan (32 GeV per beam), Cornell (5 GeV), China (2 GeV), and Russia (0.7 GeV). Additional electron-positron colliders designed for specialized investigations are under construction at Tsukuba, Stanford, and Cornell. The HERA facility at DESY is an electron-proton collider, the only one of its kind. The beam energies are 30 GeV and 800 GeV, respectively, for the electrons and protons.

Patterns and Regularities

Space-Time Symmetries

Life at the subnuclear level is complex. There are a great many different kinds of particles, and a much greater array of distinct collision and decay reactions among them. Practitioners in this field of science as in others are sustained by the belief that there must be "simplicity" underneath; and indeed, various symmetries and other patterns in the data have been identified. Particle physicists often grow rhapsodic on the theme of symmetries in nature's laws, the seemingly exact symmetries as well as certain cleverly "broken" ones. This rhapsody is well justified. Regrettably, however, it takes not just an open and romantic spirit but a considerable dose of mathematics and quantum mechanics to feel the poetry. We will have to settle here for prose.

One major set of symmetry principles comes to the subject from the past, from nineteenth-century classical physics: namely, the conservation laws of energy, momentum, and angular momentum. There was a brief scare about energy conservation in the early days of beta decay, but that has long since evaporated. Presently, there is not a shred of evidence to suggest that these three conservation laws are other than exact. Looked at in the right way, they reflect a deep set of space-time symmetry principles: for energy conservation, the notion that the basic laws of nature are unchanging with time (the same today as yesterday as tomorrow); for momentum conservation, unchanging in spatial location (the same here as there); for angular momentum, unchanging under rotation of reference frame (the same in one laboratory as in another that has a different rotational orientation). Also unchallenged are the symmetry principles embodied in special relativity, which incorporate rotation invariance and more generally require that the basic laws of nature have the same form in all inertial frames. The kinematics of special relativity are daily fare in high-energy particle physics. At a deeper theoretical level, the requirements of special relativity set a tight frame on the possible structure of quantum field theories.

In addition to the above space-time symmetry principles—invariance of the laws of nature under time translations (from one time to another), space translations (from one location to another), and Lorentz transformations (from one inertial frame to another)—there are two others of classical origin that were taken over as candidate symmetries for the microscopic, quantum world: *parity invariance* and *time reversal invariance*. The former, classically, is the assertion that the laws of physics are invariant under the simultaneous reversal of all positions and momenta, $\mathbf{r} \to -\mathbf{r}$, $\mathbf{p} \to -\mathbf{p}$. Notice that orbital angular momentum $\mathbf{L} = \mathbf{r} \times \mathbf{p}$ remains unchanged, since both $\mathbf{r}$ and $\mathbf{p}$ change sign. A classical example: Suppose that a particle moves in a time-independent, central potential $V(r)$, and suppose that $\mathbf{r}(t)$ is some particular solution of Newton's equation of motion. Then the following, which we denote with a prime, is guaranteed to be another solution (as you can easily check): $\mathbf{r}'(t) = -\mathbf{r}(t)$, hence $\mathbf{p}'(t) = m d\mathbf{r}'/dt = -\mathbf{p}(t)$. That is, the same equation that allows the one trajectory allows the other one, with position and momentum vectors each reversed in sign. Central potentials are said to be parity invariant. Time reversal invariance is the assertion that the laws of nature are invariant under a change of sign of time and momentum, position remaining unchanged. A classical example: Suppose that the potential is time independent. Then, if $\mathbf{r}(t)$ is a solution of Newton's equation, so too is $\mathbf{r}'(t) = \mathbf{r}(-t)$, and therefore $\mathbf{p}'(t) = -\mathbf{p}(t)$. Newton's equation for time-independent potentials is time reversal invariant. The classical notions of parity and time reversal symmetry, illustrated above, were taken over uncritically as hypotheses for the microscopic world in the richer context of quantum mechanics.

To illustrate the implications, first for parity invariance, consider the total cross section for a pion incident on a proton at rest. Suppose that the pion, which is spinless, is moving northward, the proton spin also pointing north. Recall that the parity operation reverses the direction of momentum but not of angular momentum vectors; hence not of spin. Thus, parity invariance implies that the cross section is unchanged if the pion motion, but not the spin, is reversed in direction, the pion moving southward, the proton spin still pointing north. But rotation invariance tells us that starting from this latter situation, there will be no change in cross section if we rotate both momentum and spin through 180°. This brings us back to a northward-moving pion, but now with proton spin pointing south. Thus, the hypothesis of parity invariance taken together with the unchallenged principle of rotational invariance tells us that the total cross section does not care which way the proton spin is pointing. Time reversal invariance in quantum mechanics is somewhat more subtle. To illustrate, consider any two-body to two-body reaction, $a + b \rightarrow c + d$. Under the time reversal operation all momenta and spins are reversed; but more striking, since we have reversed the flow of time, the direction of the arrow changes. We are now considering $c + d \rightarrow a + b$. As for the reversal of momenta and spins, that can be undone by invoking rotation invariance. Thus, the combined principles of time reversal and rotation invariance combine to relate the processes $a + b \rightarrow c + d$ and $c + d \rightarrow a + b$.

We now know that parity and time reversal invariance, though well respected in the so-called strong and electromagnetic interactions, are violated in the weak interactions.

Charge Conjugation

It is a deep symmetry principle of quantum field theory that for every particle that has electric charge, or any of several other kinds of charge we will be discussing, there is a distinct counterpart with the signs of all charges reversed but with the same mass and, if unstable, the same lifetime. One of the pair is called the particle, the other the antiparticle. Either one is spoken of as the *charge conjugate* of the other. Usually, they are denoted by the same letter, the antiparticle having a bar overhead. Thus p is the proton, $\bar{p}$ the antiproton. But there are many notational exceptions. For example, instead of e and $\bar{e}$ for electron and antielectron (positron), one usually writes e^- and e^+; similarly, π^+ and π^- for the charged pions, a particle-antiparticle pair. Particles with no charges of any kind, such as the photon γ or neutral pion π^0, are said to be their own antiparticles. They are *self-conjugate*. The notion of particle-antiparticle pairs first arose in Dirac's quantum theory of the relativistic electron. That theory, after some initial confusion, clearly entailed equality of the two masses. The ensuing development of quantum electrodynamics had automatically embedded within it a far-reaching symmetry known as *charge-conjugation* invariance. This was then absorbed as a general principle for the exploding field of subnuclear particle physics. Charge-conjugation invariance asserts that the laws of nature are invariant under the interchange of particles and antiparticles. More precisely illustrated for our purposes, the invariance principle asserts that the cross section for any collision process, or rate of any decay process, is unchanged if all the participants are replaced by their conjugates (every particle replaced by its antiparticle, every antiparticle by its particle). Thus, the processes $\pi^- + p \rightarrow \pi^0 + n$ and $\pi^+ + \bar{p} \rightarrow \pi^0 + \bar{n}$ are predicted to have the same cross sections. We have used here that π^0 remains unchanged under the switch, since it is self-conjugate.

As with the principles of parity and time reversal invariance, we now know that charge conjugation invariance, though well respected in the strong and electromagnetic interactions, is violated in the weak interactions. Indeed, parity and charge-conjugation invariance fell together in the middle 1950s, time reversal invariance somewhat less than a decade later. Importantly, although parity P, time reversal T, and charge conjugation C all break down in the weak interactions, the combined symmetry CPT remains valid. Indeed, its validity lies deeply embedded in the principles of quantum field theory. Among other things, it guarantees the equality of mass and of lifetime for particle and antiparticle.

Strong, Electromagnetic, Weak

We will come to quarks and gluons in due course; but for the present let us focus on the particles that one actually "sees" and deals with in the laboratory. Quarks and gluons leave plenty of clues, but they never actually come out clean to be seen as separable entities, at least they have not done so thus far.

The decay of the muon (mu lepton) into an electron, neutrino, and antineutrino is much slower than the similar-looking decay of a tau lepton into electron, neutrino, and antineutrino. Yet, there is a good sense in which the intrinsic propensities or strengths of these two reactions are essentially the same. The point is that the mu lepton is much lighter than the tau lepton so that there is less energy available in its decay reaction. Quite generally, whether for cross sections in the case of collision reactions or decay rates in the case of unstable particle disintegrations, the propensity for any particular reaction is the product of two factors. One, a so-called *phase-space* factor, is determined by the energy that is available for the reaction. If there is very little energy available, there just isn't much latitude for the reaction to occur. The phase-space factor is independent of details of the underlying theory and is easy to calculate. The other factor is the absolute square of a quantum mechanical quantity called the *transition amplitude*. It is this transition amplitude that provides a proper measure of the reaction's intrinsic strength. The transition amplitude depends very much on the details of the underlying theory.

Already by the middle years of the twentieth century it was recognized that particle reactions appear to organize themselves according to intrinsic strength into three distinct classes: strong, electromagnetic, and weak. What this suggested was that particle phenomena in all their great variety are rooted in an underlying foundation based on just three systems of force—much as the great variety of planetary, spaceship, and baseball trajectories are all comprehended within the one simple, gravitational force law of Newton. To be sure, there is a lot of variety in intrinsic strength within any one class. But generally speaking, electromagnetic processes have characteristically smaller transition amplitudes than comparable strong processes. At moderate energies, weak reactions are even more muted in strength, although as we will discuss later on, weak and electromagnetic strengths become comparable at very high energies. Without yet telling what the rules are, here are several examples of classification out of an almost limitless number that could have been chosen:

(i) strong: $\pi^- + p \rightarrow \Sigma^- + K^+ + \pi^0$; $\rho^+ \rightarrow \pi^+ + \pi^0$
(ii) electromagnetic: $\pi^- + p \rightarrow \Sigma^- + K^+ + \gamma$; $\pi^0 \rightarrow \gamma + \gamma$
(iii) weak: $\pi^- + p \rightarrow \Sigma^- + \pi^+ + \pi^0$; $\pi^+ \rightarrow \mu^+ + \nu$

Strictly speaking, all three of the underlying forces come into play for each and every kind of particle reaction. However, if the strong force acting alone would allow a certain reaction to proceed, that force will dominate the reaction, the other two contributing only small (though perhaps highly interesting) modifications. The process will be classified as a strong one. Next, consider a reaction that would not be allowed by the strong force acting in isolation, but suppose that it can proceed through the electromagnetic force acting alone or in combination with the strong one. In this case the electromagnetic force, serving as gatekeeper, controls the general order of magnitude of the transition amplitude. The weak force makes only minor modifications. The process is said to be an electromagnetic one. Finally, if a reaction requires intervention of the weak force, acting alone or in combination with one or both of the other forces, it is the weak one that serves as gatekeeper in setting the order of magnitude of the transition amplitude. The process is said to be a weak one.

The great majority of known particles enter into reactions of all three kinds. These particles, collectively, are known as *hadrons*. The group includes the nucleons (proton and neutron), pions ($\pi^{\pm}$, π^{0}), and much else. The hadrons are divided into two broad subgroups: *baryons* and *mesons*. Baryons are fermi particles, entities with half-odd integral spin, 1/2, 3/2, ... Mesons are bosonic particles, entities with integer spin, 0, 1, ... (we are measuring spin angular momentum in units of Planck's constant).

The particles that enter into electromagnetic and weak but not strong reactions form a smaller class. Preeminent among them is the quantum of electromagnetism, the photon. Other members of this class include the so-called weak bosons $W^{\pm}$ and the charged *leptons* $e^{\pm}$, $\mu^{\pm}$, and $\tau^{\pm}$ (electron, muon, and tau lepton).

The remaining class is composed of particles that participate *solely* in weak reactions. The neutrinos and their antiparticles belong in this collection. There are three different neutrino-antineutrino pairs: the electron neutrino ν_e, muon neutrino ν_μ, and tau neutrino, ν_τ, and their antiparticles (which would be denoted with an overbar). The neutrinos and their antiparticles are neutral members of the lepton family, whose charged members were listed just above. Another member of the weak class is the neutral weak gauge boson Z.

Cutting across different lines, the photon as well as the gluons and weak bosons W^+, W^-, and Z yet to be discussed all enter as *gauge bosons* in the modern theory. As for the quarks, we will come to them soon.

Conservation Laws, Exact and Limited

Electric charge is additively conserved; exactly, as far as we know. What additivity means in our context is that the net electric charge is the same before and after any reaction. It will be convenient here to measure charge in units of the proton charge. Using this convention, one often speaks of the electric charge quantum number of a particle. Some examples: the proton p, positive pion π^+, and positron e^+ all have electric charge quantum number +1; for the antiproton $\bar{p}$, negative pion π^-, and negative electron e^- the value is −1; for the photon, neutrinos, neutron, antineutron, and neutral pion π^0, the charge quantum number is 0. The reaction $\pi^- + p \rightarrow \pi^0 + n$ is consistent with electric charge conservation and in fact it occurs in nature. The reaction $\pi^- + p \rightarrow \pi^0 + p$ would violate charge conservation. It does not occur in nature. Before the quark hypothesis was introduced, it was generally thought that all electric charges must be integer multiples (positive, negative, zero) of the proton charge. The quarks, as it turns out, have fractional charges. We may remark here, incidentally, that the universe as a whole, so far as we know, has always been and remains electrically neutral.

Baryon number is another quantity that is additively conserved so far as we know; and what we know holds to a very high level of precision. This quantum number is nonzero *only for baryons*: its value is +1 for proton, neutron, Λ particle, charged and neutral Σ particles, and many others, and −1 for their antiparticles. The decay reaction $p \to e^+ + \pi^0$ is forbidden by the baryon number conservation law since the net baryon number to the left of the arrow is +1, but 0 to the right. Thank goodness for the taboo! Baryon conservation stabilizes the proton against this and other modes of decay that one could otherwise imagine. The mean lifetime of the proton, if it is unstable at all, is known to be not less than about 10^{33} years!

Lepton Conservation? From the kinematics of nuclear beta decay it has long been known that the electron neutrino has at most a very small mass (see Table 8.1). It seemed natural to assume that mass to be precisely zero. The muon and tau neutrino masses are less tightly restricted experimentally, but the upper limits are still small compared to the mass of the electron. When these neutrinos came along it seemed natural to assume zero mass for them also. Zero is an elegant number! As has been noted several times, there are three families of leptons: e^- and ν_e (and their antiparticles); μ^- and ν_μ (and their antiparticles); τ^- and ν_τ (and their antiparticles). In the framework of the standard model, strict neutrino masslessness has important implications. It implies the existence of three separate, additive conservation laws for leptons. For the electron family, there is the conservation law of e-type lepton number,

Table 8.1. The quarks and leptons. Especially for the lighter quarks, the masses are somewhat notional. Neutrinos have long been presumed to be massless, but there are now strong indications that some or all have nonvanishing, though tiny, masses.

Particle	*Charge*	*Mass*	*Particle*	*Charge*	*Mass*	*Particle*	*Charge*	*Mass*
u	2/3	1–5 MeV	c	2/3	1.3 GeV	t	2/3	174 GeV
d	−1/3	3–9 MeV	s	−1/3	60–170 MeV	b	−1/3	4.3 GeV
e^-	−1	0.51 MeV	μ^-	−1	106 MeV	τ^-	−1	1.78 GeV
ν_e	0	< 7 eV	ν_μ	0	< 0.17 MeV	ν_τ	0	< 18 MeV

where e^- and ν_e bear an e-type lepton number $+1$, their antiparticles have the quantum number -1, and *all* other particles have the quantum number 0. Similarly there are μ-type and τ-type quantum numbers, also strictly conserved. These conservation laws would imply, for example, that the decay $\mu \rightarrow e + \gamma$ is forbidden; and that in $\pi^+ \rightarrow \mu^+ + X$, the object X is a neutrino rather than an antineutrino; moreover, that that neutrino is ν_μ not ν_e or ν_τ.

But there is growing evidence as this is being written that neutrinos are *not* strictly massless, at least not so for all three species. The evidence is indirect and comes from fascinating quarters. One has long known that if neutrinos were to have mass there would be the theoretical *possibility* that they can shift identity among the three types as they move through space or through matter. This is the idea of *neutrino oscillations*. Thus, a ν_e produced in some decay or collision reaction might, as it moves along, convert to a quantum superposition of all three types, ν_e, ν_μ, and ν_τ, the proportions oscillating back and forth over time. The rate of oscillation would depend on differences among the neutrino masses, on the energy, and on various "mixing" parameters. None of this is mandated, but it was known to be a theoretical possibility if neutrinos have mass. On the experimental side, evidence for neutrino oscillations has been developing on several fronts. For one thing, there appears to be a shortage of electron-type neutrinos coming to us from the sun. This is the only neutrino type produced in the burning sun. The observed flux seems too small, roughly by a factor of two. Conceivably, the solar models that predict the neutrino flux are in error. However, the growing opinion is that the shortage is real, as if some portion of the e-type neutrinos have oscillated into the other neutrino species on the way to earth. There is a related anomaly, recently established on rather firm experimental grounds. This has to do with fluxes of e-type and μ-type neutrinos that are generated in the cosmic ray atmosphere and that reach detectors placed underground. Again, there are shortages, this time in the abundance of μ-type relative to e-type neutrinos. It looks as if there are oscillations going on here too!

The implications of all of this are under intense scrutiny as this is being written and it may be best to bow out here after a few more lines. It seems probable that the three separate lepton conservation laws are in for a fall, though it may be that the violations will be quite small. It is still possible, however, that a single, overall lepton conservation law can survive: all three neutrinos and three negatively charged leptons having overall lepton quantum number +1; their antiparticles, −1; everything else, 0. In the following, when we speak of lepton conservation we will be referring to this overall quantum number. Another point of enormous interest: The universe is populated with neutrinos left over along with photons from the Big Bang. It has been known for some time that the universe is filled with some sort of energy content that makes itself felt gravitationally but does not otherwise manifest itself. This is the cosmological "missing mass" problem (mass being equivalent to energy according to Einstein). If neutrinos have mass, even as small as a few electron volts, they could contribute a significant portion of the missing mass of the universe.

In contrast to the seemingly exact conservation laws for electric charge, baryon number, and, possibly, overall lepton number, there are other quantities that were known from the moment of their introduction to be conserved in only a limited domain. They are additively conserved in strong and electromagnetic reactions but invalidated in the weak ones. Altogether, there are four such quantities. *Strangeness* number is one of these. The concept arose in the 1950s with the discovery that hadrons known to participate in strong and electromagnetic reactions in certain combinations react only weakly in other combinations. We saw some examples earlier on. One could accommodate this by assigning a new kind of quantum number, strangeness, to the various hadrons in such a way that strangeness is additively conserved in strong and electromagnetic but not in weak reactions. Even in the weak reactions there are patterns to the breakdown of the conservation law. In ordinary weak processes, strangeness changes by only one unit between the two sides of the equation. Reactions in which it changes by more than one unit are not strictly forbidden but they are *very* weak (and are the objects of intense experimental search). To illustrate: Nucleons and pions all have strangeness $S = 0$, whereas the K^+ meson has strangeness $S = 1$. Accordingly, the reactions

$$\pi^+ + p \to \pi^+ + \pi^+ + n,$$

$$\pi^+ + p \to K^+ + \pi^+ + n,$$

$$\pi^+ + p \to K^+ + K^+ + n$$

are, respectively, strong, ordinary weak, very weak.

There are three other quantities analogous to strangeness that are additively conserved in strong and electromagnetic but not in weak reactions. They came along later, after the quark hypothesis had taken hold. These three, taken together with strangeness, baryon number, and electric charge, constitute a set of six additive conservation laws for the strong and electromagnetic reactions. Six is the number of quark types! Although these conservation laws find a natural home in modern, quark-based theory, it must be emphasized that they can be read off directly from experimental data without reference to any underlying quark theory.

To the Quarks

A number of other approximate symmetry ideas have been proposed and supported by the data. The concepts that are involved are somewhat more complicated than for the additive conservation laws. One example is *isotopic spin* symmetry, which holds quite accurately in strong reactions. Here one assembles hadrons into groups, or isotopic multiplets as they are called. The members of a given multiplet all have the same spin, baryon, and other additive quantum numbers except for electric charge. To the extent that symmetry-violating effects arising from electromagnetic and weak forces can be ignored, all the members of a given multiplet should have exactly the same mass. Thus (p, n) form the nucleon doublet, (π^+, π^0, π^-) constitute the pion triplet, (Λ) is the lambda particle singlet, (K^+, K^0) the K meson doublet; and so on for groupings of other hadrons. It is already clear that there is merit to the isotopic spin grouping just from the fact that p and n *do* in fact have very nearly the same mass, that the neutral pion *does* have almost the same mass as the charged pions, and so on for the other multiplets. Isotopic spin symmetry goes beyond this, however. It is often powerful enough to predict relations among the cross sections of various processes involving a given set of multiplets. For example, it yields definite relations (which we do not bother to write down) connecting the cross sections for the processes

$$\pi^- + p \to \Lambda + K^0, \quad \pi^0 + p \to \Lambda + K^+,$$

$$\pi^0 + n \to \Lambda + K^0, \quad \pi^+ + n \to \Lambda + K^+.$$

In general, isotopic spin predictions are well confirmed by the experimental data.

By the very early 1960s another, more inclusive symmetry of the strong reactions was proposed. One knew from the start that it was inexact, but it was pursued nonetheless as a possibly useful approximation. This is SU(3) symmetry, a mathematical terminology that we need not go into here. It groups various isotopic spin multiplets together into larger multiplets, all the particles in a given multiplet having the same spin and baryon number. If the symmetry were exact they would in addition all have the same mass. For example, the pion isotopic triplet, the K meson doublet, the anti-K doublet, and the η particle singlet, all with the same spin and same vanishing baryon number, are collected together in a single SU(3) multiplet with 8 members. Other hadrons are similarly grouped together in other multiplets, with *dimensions* (number of members) allowed by SU(3) symmetry; for example 8, 10, 27. Alas, the masses within the various SU(3) multiplets are not all the same, far from it in some cases; so SU(3) symmetry is hardly exact. Nevertheless, it provides a reasonable approximation in many situations.

But the chief triumph for SU(3) symmetry was its role in generating the quark hypothesis. The mathematics of SU(3) allows for multiplets of dimension 3. After some early false starts, it became clear that none of the known hadrons could sensibly be grouped into multiplets of this dimension; they all had other homes. This by no means constituted a contradiction of any sort. One could just say that nature, for her own reasons, chose to ignore a simple possibility afforded by the mathematics of SU(3). Nevertheless, the idea of a substructure of hadrons based on an SU(3) triplet of entities, quarks, began to take hold in the early 1960s. Although the underlying quantum dynamics was far from clear, at least the mathematics of SU(3) symmetry allowed one to picture the then known hadrons as being made up of combinations of three types of (presumably) spin-one-half quarks (here we use "quark" in a collective sense to include both particle and antiparticle). Those first three quarks have come to be called *up*, *down*, and *strange* (one must steel oneself for the excesses of whimsy in subnuclear nomenclature). They are symbolized respectively by the letters u, d, and s.

In the early stages, the quarks were regarded by many as mere mathematical crutches, to be discarded when they had provided their mathematical shortcuts and clues. For others they were real physical particles, to be searched for experimentally. What we now believe is located somewhere in between. The quarks are very real indeed in the sense that they enter as basic ingredients in the modern theory of particles. They leave their unmistakable fingerprints in the right kinds of experiments. But they seem never to emerge directly, to be inspected in isolation.

Basic Ingredients

The Particles

Those first three quarks were followed over subsequent years by the successive (always somewhat indirect) discovery of *charm*, *bottom*, and *top* quarks, denoted by the letters c, b, and t. The first of these was a prediction of the electroweak unification theory that had been introduced in the late 1960s. The discovery of charm a few years later was one of several dramatic confirmations of the theory that occurred around that time. Then, at least for true believers, the discovery of the totally unforeseen tau lepton in the mid 1970s implied the existence of two additional quarks. Sure enough, bottom showed up in a few years' time; top took nearly two more decades. Altogether, there are six types of quark; or, as one says, six quark *flavors*. But more contemporary theory says that each quark flavor comes in three distinct varieties or subspecies, all three with the same mass, charge, baryon number, and spin. What distinguishes one subspecies from another is well defined mathematically in the context of the underlying theory, but for everyday usage one needs everyday names. In keeping with the pattern of whimsy established by the names of the quark flavors, the subspecies are designated by the names of colors. Any three colors will do, this is only a matter of nomenclature. We may use red, white, and blue. In the following discussion, we will simply speak of the six quark flavors, with the understanding that for each flavor there is both a particle and an antiparticle, each coming in three colors; hence, really, 36 different entities. The quarks are spin $= 1/2$ fermions, a feature that they share with the three charged leptons and three neutral leptons (neutrinos).

The quarks have fractional electrical charges. In units of the proton charge, the charge quantum numbers for the quark *particles* are $Q = 2/3$ for u, c, t; $Q = -1/3$ for d, s, b. For the quark *antiparticles* (antiquarks), the signs are exactly reversed: $Q = -2/3$ for $\bar{u}$, $\bar{c}$, $\bar{t}$; $Q = 1/3$ for $\bar{d}$, $\bar{s}$, $\bar{b}$. Similarly, the baryon number for all six quark particles is $B = 1/3$; for the quark antiparticles the sign is reversed, $B = -1/3$. In themselves these fractional values are not mystifying. If, for example, the down quark had been discovered before the other quarks and before electrons and protons, it might have become the standard relative to which other charges are measured; the electron, when finally discovered, would then acquire the charge quantum number 3. But of course that's not the way it happened historically; and it came somewhat as an initial surprise that there could be entities with charges smaller in magnitude than the familiar electron or proton charge.

Because quarks never emerge in isolation, it is hard to determine their masses with great precision; indeed, there is some theoretical looseness about how the mass parameter should be defined. But we know well enough that the up and down quark masses are quite small on the scale of familiar hadron masses; and that the strange quark mass is somewhat larger, though still fairly modest on that scale. Charm, bottom, and top quarks have masses that are very much larger than those of the three lighter quarks, and their masses can be determined fairly accurately. Organized into three *families*, Table 8.1 lists the quark and lepton particles along with their masses and electric charges. Not indicated are the (baryon, lepton) numbers, which are $(1/3, 0)$ for the quarks and $(0, 1)$ for the leptons. The corresponding *antiparticles* need not be separately listed; they have the same masses but reversed electric charge, baryon, and lepton numbers.

The modern theory also introduces a set of other particles, *gluons*. Just like the quarks, they leave their unmistakable fingerprints in the right experiments but they never come out in isolation. There are 8 of them. They are massless and electrically neutral, with baryon number zero. They play the same kind of role in strong interactions that the photon γ does in electromagnetic interactions and the W^+, W^-, and Z bosons do in weak interactions. All are spin-one *gauge* bosons, a total, then, of $8 + 4 = 12$ of them. The photon, gluons, and Z particle are their own antiparticles, whereas W^- and W^+ are a charge-conjugate pair. The gauge bosons are listed in Table 8.2.

Altogether, our list of basic ingredients consists of six quark flavors, six lepton flavors, and twelve gauge bosons; but recall

Table 8.2. The gauge particle: photon, gluons, charged and neutral weak bosons

γ	g	W^+, W^-	Z
0	0	80 GeV	91 GeV

that each quark flavor comes in three colors and that for quarks and leptons there are both particles and distinct antiparticles. In the minimal contemporary theory, the standard model, there is only one other particle to be added to the list—the Higgs particle, a neutral, spin-zero object. At this writing it has not yet been discovered but is being intensely hunted. It has a central role to play in that it is thought to be the source of particle masses. But that role is complicated and we will drop the Higgs particle from further discussion here.

The lists in Tables 8.1 and 8.2 may some day broaden. Indeed, there is intense, current speculation about possible *supersymmetric* partners to all the particles in our collection, and about other extensions of the minimal picture. But what is in any case conspicuously missing from the tables of fundamental ingredients are the proton and neutron, the pions, and all the other hadrons. On the modern view, they are composites formed out of quarks and gluons. The atomic nuclei of "everyday" life are therefore composites of composites. Moreover, the quarks and gluons, although they presently rank as fundamental building blocks, are in a sense phantoms; they never come out into the open. That leaves only the leptons, weak gauge bosons, and photon as both fundamental (according to contemporary theory) and *directly* accessible.

The Interactions

A comprehensive theory must specify not only the fundamental particles that it builds on, but also the forces that govern their behavior. However, in a world of particle creation and destruction it is better to speak not of forces but of fundamental *interactions*, the elemental acts of creation and destruction that combine to produce collision and decay reactions and that determine the structure of composites such as the hadrons. Later on, we will try to spell out the notion of fundamental interaction. For the moment, let us proceed loosely.

Strong

The strong interactions involve a sum of flavor-preserving interaction terms, one for each quark flavor. Each term describes the coupling of a gluon g to a pair of quarks q or pair of antiquarks *of the same flavor*, or to a quark-antiquark *pair of the same flavor*. We symbolize these by $q+g \leftrightarrow q$. Here and in what follows, the one equation of this sort is meant to include also $\bar{q}+g \leftrightarrow \bar{q}$, $q+\bar{q} \leftrightarrow g$, and $q+\bar{q}+g \leftrightarrow 0$. The symbol 0 stands for "nothing"—the no-particle state. Notice that when any entity moves from one side of the double-headed arrow to the other, it converts to its charge conjugate. A measure of the strength of these couplings is embodied in a parameter called the *strong coupling constant*. Its value is the same for all six quark flavors. In a notional sense we take the strength to be of order unity. The electromagnetic and weak couplings are markedly smaller than unity.

More complicated reactions among quarks and gluons arise from these fundamental ones and certain others involving interactions purely among the gluons. Later on we will illustrate how these more complicated reactions get built up out of the fundamental interactions. For the present, the important thing to notice is that the fundamental couplings described above are flavor preserving; that is, an X quark or antiquark remains the same X quark or antiquark after absorbing or shaking off a gluon; similarly, an X quark can annihilate only against an antiquark of the same flavor to produce a gluon. Here X is any one of the six quark flavors. Another way to say this is that the number of quarks of a given flavor minus the number of antiquarks of the same flavor is the same on both sides of any of these fundamental interaction equations. The same must then be true of more complicated reactions built up out of these fundamental ones. *The strong interactions are flavor conserving*. That is, the theory implies the existence of six additive conservation laws for strong reactions. One of these is a conservation law for N_u, the number of up minus the number of anti–up quarks; another is for N_d, the number of down minus the number of anti–down quarks; and so on. Any combination of these conserved quantities is, of course, also a conserved quantity. We recognize thereby the conservation of baryon number N_B and electric charge number N_Q within the framework of strong reactions, with

$$N_B = \frac{1}{3}(N_u + N_d + N_c + N_s + N_t + N_b),$$

$$N_Q = \frac{2}{3}(N_u + N_c + N_t) - \frac{1}{3}(N_d + N_s + N_b).$$

Except for small contributions arising from electromagnetic and weak interaction effects, hadrons are made of quark particles, quark antiparticles, and gluons. The proton, for example, has the conserved quark numbers $N_u = 2$, $N_d = 1$, all the other quark quantum numbers vanishing ($N_s = N_c = \cdots = 0$). This gives, correctly, $N_B = 1$, $N_Q = 1$. The simplest interpretation would then be that the proton is literally made up of two u quarks, one d quark, and nothing else. But that is surely too naive. So far as quantum numbers are concerned you can add any number of gluons to the mix since they carry no charge or baryon number. Similarly, you can add any number of quark-antiquark *pairs* of any flavor without changing the proton's quantum numbers. These constitute a so-called "sea" of quark-antiquark pairs, presumably the lighter up, down, and strange quark pairs predominating. Certainly the proton contains 2 more u-flavored quarks than antiquarks, 1 more d-flavored quark than antiquark. But quantum numbers do not carry us beyond that, or tell us anything about the sea or the gluon content of the proton. These are deeper, more detailed questions that have to be addressed within the underlying dynamical theory. They are the very difficult subjects of ongoing theoretical and numerical analysis. Having said this, at least for quantum number characterization we can describe the proton with the self-evident notation (uud), meaning $N_u = 2$, $N_d = 1$, all other Ns $= 0$. The antiproton is then ($\bar{u}\bar{u}\bar{d}$). In the same quantum number sense, the positive pion π^+ is the quark-antiquark combination ($u\bar{d}$); its antiparticle π^- is ($d\bar{u}$). The neutral pion π^0 is the linear combination ($u\bar{u} - d\bar{d}$). You can tell that particle and antiparticle are one and the same for the neutral pion since the composition is unchanged if each quark is replaced by its antiquark and vice versa.

Table 8.3 lays out a very small sample of the known hadrons. It lists their masses and quark characterization. The baryons on the list all have baryon number $B = 1$; the mesons, $B = 0$. The former necessarily have distinct antiparticles. As for the mesons, some are their own antiparticles. They can be recognized by applying the test described above for the neutral pion (does the quark content change under charge conjugation?). It should be emphasized that many different hadrons can have the same quark characterization. For example, there is a whole array of baryons with the proton's structure (uud), all differing in mass and other characteristics.

A final comment here. We had earlier touched briefly on the isotopic spin and SU(3) symmetries of strong reactions. Ignoring the small contributions of electromagnetic and weak interactions, isotopic symmetry would be exact for strong reactions if the up and down quark masses were identical. In fact they are not numerically identical, but both are very small compared to typical hadron masses. They are approximately identical, then, in the sense that the u and d masses can equally be regarded as negligible in many contexts. The more inclusive SU(3) symmetry would be exact if the masses of the three quarks u, d, and s were all equal. In fact, the s-quark mass differs substan-

Table 8.3. A small selection of the strongly interacting particle, hadrons

Baryons	*Quark structure*	*Mass MeV*	*Mesons*	*Quark structure*	*Mass MeV*
p	uud	938	π^+	$u\bar{d}$	140
Λ^0	uds	1116	K^0	$d\bar{s}$	498
Δ^{++}	uuu	1232	D^0	$c\bar{u}$	1865
Ξ^0	uss	1315	D_{S+}	$c\bar{s}$	1969
Ω^-	sss	1672	J/Ψ	$c\bar{c}$	3097
Λ_{c+}	udc	2285	B^+	$u\bar{b}$	5279
Ξ_c^0	dsc	2470	B_s^0	$s\bar{b}$	5370
Λ_b^0	udb	5624	Y	$b\bar{b}$	9460

tially from that of u or d and is not all that negligible; so the symmetry is at best only roughly approximate.

Electromagnetic

The electromagnetic interactions are described by a sum of terms coupling a photon γ in turn to every charged particle Q on our lists: $Q + \gamma \leftrightarrow Q$. As before, we understand this to include $Q^{\pm} + \gamma \leftrightarrow Q^{\pm}$, $Q^{+} + Q^{-} \leftrightarrow \gamma$, and $Q^{+} + Q^{-} + \gamma \leftrightarrow 0$. The characteristic coupling constant for any of these interactions is the electrical charge of the particle. It is smaller than the strong interaction coupling constant. Since the u and d quarks have different charges, electromagnetic interactions violate isotopic spin symmetry. However, they conserve flavor and hence baryon and lepton number; and, of course, they conserve electric charge.

The strong interactions are mediated, as one says, by the gluons, which couple to pairs of quarks of the same flavor. Electromagnetic interactions are mediated by the photon, which also couples to pairs of quarks of the same flavor, to pairs of charged leptons of the same flavor, and to the charged W bosons. Taken together, the strong and electromagnetic interactions conserve flavor for the quarks and lepton number for each of the three species of charged leptons. The neutrinos have not yet come into play.

Weak

The weak interactions are mediated by the weak gauge bosons, W^{+}, Z, and W^{-}. The neutral boson Z couples to pairs of quarks and pairs of charged leptons in much the same way that the photon does; in particular, the couplings are flavor preserving for the quarks and charged leptons. Recall what this means. It means that a u quark interacting with the Z boson remains a u quark, a d quark remains a d quark, an electron remains an electron, and so on. What is qualitatively new is that neutrinos now come into play. The Z couples to pairs of neutrinos of the same flavor. As for the W bosons, they couple to pairs of quarks with different flavors, necessarily so in order that charge be conserved. They also couple to pairs of leptons, one charged, the other a neutrino. Altogether then, apart from interactions coupling the gauge bosons among themselves, the couplings of the W bosons are as follows:

$$W^+ \leftrightarrow u + \bar{d}, \quad c + \bar{s}, \quad t + \bar{b}; \quad e^+ + \nu_e, \quad \mu^+ + \nu_\mu, \quad \tau^+ + \nu_\tau.$$

We are again using shorthand here. Each reaction symbolizes itself and others. For example, $W^+ \leftrightarrow u+\bar{d}$ includes $\bar{u}+W^+ \leftrightarrow d$, $d + W^+ \leftrightarrow u$, $W^- \leftrightarrow \bar{u} + d$, and so on.

The distinctive features of the weak interactions are that they bring in the neutrinos; and that they generate flavor-changing transitions among the quarks and hence among the hadrons.

The coupling constants in the above weak interactions are all of the same order of magnitude as the electromagnetic coupling constants. This reflects one of the great triumphs of the modern theory; namely, the unification of electromagnetic and weak interactions. Although the electromagnetic and weak coupling constants are roughly of the same strength, weak reaction transition amplitudes are far smaller than electromagnetic ones at low energies. "Low energy" here means small compared to the very large rest mass energies of the weak gauge bosons. As we will discuss in the next chapter, this comes about because at low energies the very large masses of the weak gauge bosons appear in denominators and tend to suppress transition amplitudes.

Summary

As we now think, the world is built on a foundation of six flavors of quarks, three charged leptons and their neutrinos, and gauge bosons for each class of fundamental interaction: eight gluons for the strong interactions, a single photon for the electromagnetic interactions, the three "weak" bosons, W^+, Z, W^- for the weak interactions. It should be said again that we are speaking here of quarks in a collective sense to include both particle and antiparticle; so too for the leptons. Still undiscovered is an expected neutral, spinless boson, the Higgs particle.

Not on our lists, most strikingly, are the proton, neutron, pions, and other hadrons, even though these constitute the bulk of known subnuclear particles. They are composites, made of quarks and gluons.

The modern theory stands on two legs, a strong interaction component (quantum chromodynamics, QCD) and a unified electroweak component. We have not examined the theory in any detail beyond indicating its particle and interaction ingredients and noting some of its exact and limited symmetries. A closer look would reveal a deeper gauge symmetry structure of the theory, but that would quickly carry us far afield into highly technical thickets. To date, the standard model has passed all experimental tests except for the above-noted rumblings on the neutrino front—rumblings that can, however, be accommodated without great violence. Even so, there are various reasons to think that the present theory must be embedded in some wider-reaching framework, that it is not complete. For one thing, it contains uncomfortably many input parameters, about a dozen and a half of them. Among these are the various masses. What is especially vexing about the masses is the enormous range that they span, from the tiny electron mass to the much larger tau lepton mass, from the small up and down quark masses to the enormous top quark mass, and somewhere on the very low end of the scale, possible small but nonvanishing neutrino masses. Why? Moreover, gravity is not included in the standard model.

CHAPTER NINE

Quantum Fields

The subnuclear particles that we are concerned with are tiny things that leave tracks in detectors of various sorts, or trigger Geiger counters, or register themselves in other particle-like ways. If stable they have definite masses; if unstable, definite lifetimes and almost definite masses. A certain subset of them, electrons, protons, and neutrons, combine in large numbers and in various groupings to make up the material, macroscopic world of everyday life. Photons, taken together in large numbers, make up the everyday world of light (and radio waves and X-rays and so on). For all these reasons, at the microscopic level it is clearly the particle aspect of the world that attracts our interest. Nevertheless, from a modern point of view particles are not the primary theoretical constructs. That honor goes to quantum fields. Let us again recall the classical situation.

Classically, particles and fields are dynamical quantities of equal status. Any given particle is at a definite position at each instant of time. The dynamical goal is to predict how that position varies with time. The time variation is governed by Newton and by the relevant force laws. In contrast, a classical field $\phi(x, y, z, t)$ is a quantity defined continuously over all space. The dynamical goal is to predict how the field varies with time *at each position of space*. Since there is a continuous infinity of points of space, there are infinitely many dynamical variables, or degrees of freedom. The dynamics are governed by appropriate partial differential equations, for example Maxwell's equations for the set of electric and magnetic fields. A general dynamical system will contain both particles and fields: for electromagnetism, charged particles as well as **E** and **B** fields.

How does one go about quantizing fields? First, recall how this is done for a system of particles. There the basic classical observables are the position and momentum vectors of the particles. Other quantities of interest such as the energy of the system, its angular momentum, and so on, can be expressed in terms of these. With that done, one now quantizes by converting the position and momentum observables into operators (which we denote with an overhead tilde). In the Schroedinger representation that we have adopted, these operators are time independent. Other observables such as energy now also become operators. The state of the system is encoded in a wave function whose time evolution is governed by the Schroedinger equation (4.19). Basic to all of this are the commutation relations among the position and momentum operators. The position or momentum operator of any one particle commutes with the position and momentum operators of all the other particles. For any given particle, the only nonvanishing commutators are

$$[\tilde{x}, \tilde{p}_x] = [\tilde{y}, \tilde{p}_y] = [\tilde{z}, \tilde{p}_z] = i\hbar. \tag{9.1}$$

These commutation relations taken together with the Schroedinger equation are at the heart of quantization for a system of nonrelativistic particles.

Free Fields, Free Particles

Analogous procedures suggested themselves early on for quantization of the electromagnetic field. This is a field system that we encounter classically; but other fields—ones that don't show up classically—have been invented over subsequent years precisely for purposes of quantization. We will be interested only in fields obeying relativistically invariant equations and begin here with a simple model, which we introduce only for pedagogical purposes: a single, scalar field $\phi(x, y, z, t)$ that at the classical level obeys the following differential equation:

$$\frac{1}{c^2}\frac{\partial^2\phi}{\partial t^2} - \left\{\frac{\partial^2\phi}{\partial x^2} + \frac{\partial^2\phi}{\partial y^2} + \frac{\partial^2\phi}{\partial z^2}\right\} + \rho^2\phi = 0. \tag{9.2}$$

The constant ρ will acquire a physical interpretation later on; for the present it is just a parameter. It is easy from Eq. (9.2) to discover a quantity that is unchanging with time and that can be identified as the energy content of the field. Up to a multiplicative constant that depends on conventions, the energy *density* (energy per unit volume) is

$$H = \frac{1}{2c^2}\left\{\left(\frac{\partial\phi}{c\partial t}\right)^2 + \left(\frac{\partial\phi}{\partial x}\right)^2 + \left(\frac{\partial\phi}{\partial y}\right)^2 + \left(\frac{\partial\phi}{\partial z}\right)^2 + \rho^2\phi^2\right\}. \tag{9.3}$$

We will refer to this as the Hamiltonian density. There are similar expressions for the momentum and angular momentum densities carried by the field.

For guidance on quantization, let us return to particle dynamics and consider a single particle of mass m moving in a potential $V(x, y, z)$. The analog of Eq. (9.2) is the Newton set

$$\frac{\partial p_x}{\partial t} = -\frac{\partial V}{\partial x}, \quad \frac{\partial p_y}{\partial t} = -\frac{\partial V}{\partial y}, \quad \frac{\partial p_z}{\partial t} = -\frac{\partial V}{\partial z}; \qquad \mathbf{p} = m\frac{d\mathbf{r}}{dt}.$$

The analog of Eq. (9.3) is the the total energy, or Hamiltonian:

$$H = \frac{1}{2m}(p_x^2 + p_y^2 + p_z^2) + V.$$

There are three position variables, x, y, z, and three corresponding momentum variables, p_x, p_y, p_z. In Eq. (9.3), the counterpart of the three position variables is the infinite set of time-dependent field variables $\phi(x, y, z, t)$, one for each point in space. Recalling that the momentum components of the particle are proportional to the time derivative of the corresponding position coordinates, we may think of the time derivative of the field $d\phi/dt \equiv \pi(x, y, z, t)$ as being the "momentum" variable that corresponds to $\phi(x, y, z, t)$. This suggests the following. To quantize, let $\phi \to \tilde{\phi}(x, y, z)$ and $\pi \to \tilde{\pi}(x, y, z)$ be time-independent but position-dependent operators. Notice that the argument (x, y, z) of these operators is not itself an operator; it is merely a label that specifies location in space, each point in space having its own field operator. The idea now is to impose commutation relations analogous to those in Eq. (9.1); namely, to require that all commutators among these operators vanish, independent of whether the two operators are at the same or different points, except for the commutator $[\tilde{\phi}(\mathbf{r}), \tilde{\pi}(\mathbf{r}')]$. By analogy with Eq. (9.1) one might think that this should be zero if the two space points are different, equal to $i\hbar$ if the points are the same. More properly, however, since space is a continuum, consider the commutator $[\tilde{\phi}(\mathbf{r}), \tilde{\pi}(\mathbf{r}')]$, fix $\mathbf{r}$, and integrate the position variable $\mathbf{r}'$ over an infinitesimal volume surrounding the point $\mathbf{r}$. A better analogy with Eq. (9.1) that suggests itself is that it is the resulting integral that should be set equal to $i\hbar$.

The total energy and total momentum observables are now represented by operators. They are integrals over space of corresponding densities expressible in terms of the basic operators $\tilde{\phi}(\mathbf{r})$ and $\tilde{\pi}(\mathbf{r})$; and we know the basic commutation relations. Done! This is all that's needed to address the eigenvalue problems for energy and momentum. Energy and momentum are commuting observables in this model, as must be the case for any realistic theory. Consequently, there are simultaneous eigenstates of these two observables. We have started with a classical field equation so simple that the quantized version is also simple. The eigenvalue problem is easily solved. The results are quite remarkable, as follows.

(1) There is a unique state of zero energy and momentum. It is the so-called vacuum state. It is the state of nothingness.

(2) The allowed momentum eigenvalues $\mathbf{p}$ form a continuum, all magnitudes and directions being allowed. For any given momentum $\mathbf{p}$ there is a particular state with energy

$$E = \sqrt{(cp)^2 + (mc^2)^2}, \qquad \text{where } m = \hbar\rho/c.$$

This is just exactly the relativistic energy-momentum relation that holds for a material particle of mass m. It is natural to interpret this state as in fact describing just that; we may speak of this as a one-particle state. A *particle* has somehow emerged out of the quantum *field*. The parameter ρ that we started with fixes its mass.

(3) There is a family of states of momentum $\mathbf{p} = \mathbf{p}_1 + \mathbf{p}_2$ with energy $E = E_1 + E_2$, the energies E_1 and E_2 being related, respectively, to $\mathbf{p}_1$ and $\mathbf{p}_2$ as in item 2 above. This is clearly a family of two-particle states labeled by the momenta $\mathbf{p}_1$ and $\mathbf{p}_2$.

(4) And so on. There are states with all possible numbers of particles, each particle with its own momentum and related energy, the total momentum and total energy being sums of the contributions from the individual particles.

Consider what has been wrought. The model field theory turns out to describe particles, all possible numbers of them. In the quantum theory treated in the earlier chapters we *started* with (nonrelativistic) particles, the number of particles in any given system being a prescribed quantity. Let us refer to that theory as quantum *particle mechanics,* in contrast to quantum *field* theory. In the model field theory we don't start with particles at all. They emerge on their own as quanta of the field; and particle number is now an observable having various possible outcomes. More striking still, in multiparticle states all the particles are *exactly* identical. They have the same mass, and in our example the same zero spin. In quantum particle mechanics there is nothing in the theory to rule out a world in which identical particles *don't* exist; for example, there is nothing that rules out a world in which all the things we call electrons are subtly different one from another. But in the model field theory there are states with all possible numbers of particles and the particles are exactly identical. We have no choice in the matter. In theories involving several different kinds of fields there may be several different species of particles. But again, there are states with all possible numbers of each kind of particle, and all the members of a given species are exactly the same.

The big trouble with our model theory is that it is dull. It describes a hypothetical world in which nothing interesting happens! Start out with a state in which two particles approach each other as if for a collision. In fact they won't collide. They will pass right by one another. This reflects the fact that the classical field equation (9.2) on which the quantum model is based is linear: the sum of any set of solutions is also a solution. This is an example of what is called a *free field theory*, a theory free of interactions. The particular theory we have discussed here happens to describe neutral, spin-zero bosons, but it is easy enough to construct analogous linear theories for charged as well as neutral particles of various spins. Theories involving charge yield both particles and antiparticles as their quanta. Altogether then, at the free field level it is easy enough to set up a multifield theory whose quanta include all the species we may think to be fundamental in the real world—leptons, quarks, and so on. But nothing happens. This is the same situation we meet in quantum particle mechanics. There, particles move freely and independently if there are no forces among them. If there is to be any action, there have to be forces. The analog for quantum field theory is that there have to be field *interactions*: mathematically speaking, nonlinear terms in the differential equations of the theory.

Interactions

The general formalism of quantum particle mechanics is relatively indifferent to the nature of the forces acting on the particles. Of course, different force laws can have very different physical consequences, some force laws may be easier to handle mathematically than others, some may be very much farther from physical reality than others; but unless a force law is pathological, the theory will at least be self-consistent. In quantum field theory it can be quite otherwise. Randomly chosen interaction terms, even innocent-looking ones, not only might be unrealistic physically, they might incorporate internal inconsistencies and other pathologies. Relativistic quantum field theory is very constraining and demanding. That's good. Quantum field theory is also very difficult mathematically. That's bad. There are no remotely realistic theories that are exactly soluble.

Having absorbed these disclaimers, let us proceed with our model field theory, adding to it a simple interaction term. Add a term proportional to ϕ^3 to the left side of Eq. (9.2). We have thereby added the following interaction term to the Hamiltonian density in Eq. (9.3):

$$H_{\text{int}} = \lambda\phi^4. \tag{9.4}$$

The above proportionality factor shows up here as the "coupling constant" λ. Let us consider what the consequences might be, at least as analyzed by the techniques of perturbation theory that we will be discussing a bit later on. As it happens, for the particular model under discussion there is serious mathematical doubt whether the above modification really produces a self-consistent theory of interacting particles. But here we sweep all such delicacies under the rug. The model is designed only for pedagogical purposes and should serve to bring out features, at least in a perturbation-theoretic sense, that can be expected to hold for the more realistic theories that will be taken up later on.

The time development of any quantum system is governed by the Schroedinger equation (4.19), which is controlled by the Hamiltonian of the system. In the absence of the interaction term the model theory yields its quanta. But they don't do anything. It is the interaction term in the Hamiltonian that will make things happen. It induces a host of scattering reactions, limited in number only by energy-momentum conservation. When two particles (we will call them mesons) collide at any energy, however small, there will be elastic scattering—two incoming, two outgoing mesons. At higher energies, there will also be events with four outgoing mesons, and so on; more and more channels opening up beyond limit with increasing collision energy. In this particular model it happens that the cross section for producing an odd number of outgoing particles is exactly zero. That's so because the total number of particles involved in any reaction, incoming plus outgoing, has to be even. As we will see shortly, this in turn follows from the fact that H_{int} is an even polynomial in the field ϕ.

Recall that the cross section for any particular reaction is the square of a transition amplitude multiplied by an easily computed phase-space factor. The heart of the matter is the transition amplitude. Exact calculations are hopelessly out of the question in our present era, so various approximation procedures have to be resorted to. Among these, the so-called perturbation approach is the one that lends itself best to intuitive description. In the context of our model theory, the idea is to imagine expanding any wanted transition amplitude as a power series in the coupling constant λ. For example, the transition amplitude for elastic scattering is a function of λ as well as of the collision energy and scattering angle. Expanded in powers of λ, the amplitude is an infinite sum of terms, each depending on energy and angle. The leading term is proportional to λ^1, the next term to λ^2, and so on. There are quite definite mathematical rules for computing each term in the series, although the calculational demands grow ferocious with increasing order (increasing powers of λ). Moreover, even if it is assumed that the series converges, for an exact answer an infinite number of terms in the expansion would have to be computed and summed. The perturbation approach is quantitatively useful, therefore, only if the coupling constant is sufficiently small so that the first few terms in the power series expansion provide a good enough approximation. This is the case for many electromagnetic and weak reactions that we will be coming to soon. Quantitatively, the perturbation approach is of more limited use for strong reactions, where the coupling constant is much larger.

The classical field of our model has become the space-dependent operator $\tilde{\phi}$. It can both create and destroy a meson;

that is, acting on a state containing n mesons it produces a new state that is a linear combination of states with $n+1$ and $n-1$ mesons. The interaction term H_{int} acting on a state with a given number of mesons can therefore create four additional mesons; destroy four mesons; create three, destroy one; create one, destroy three; create two, destroy two. These interactions are represented collectively by the diagram at the left of Fig. 9.1, which shows four lines joined at an interaction *vertex*. Any particular one of the interactions noted above can be singled out by the use of arrows, an arrow pointed *toward* the vertex denoting *destruction* of a meson, an arrow directed *away* from the vertex denoting *creation*. For example, the three diagrams to the right of the equals sign in Fig. 9.1 illustrate this, respectively, for the $2 \to 2$, $1 \to 3$, and $3 \to 1$ meson transitions.

Of course, one meson can't really, physically, convert to three mesons; or the other way around either. Neither can four mesons appear out of the vacuum, or the other way around. Energy-momentum conservation forbids these things. For example, in the case of the $1 \to 3$ transition, imagine sitting in the rest frame of the initial meson. The net momentum there is zero. The outgoing three mesons must therefore have momenta that add up vectorially to zero. That's OK. But energy will not be conserved, since the initial energy is just the rest energy of the initial meson whereas the final energy cannot be less than three times as big. These processes therefore represent only *potentialities*, tendencies that are however foreclosed by energy-momentum conservation as actual physical processes. In a sense that will be described shortly, the potentialities get to be realized in transitions involving *virtual* particles.

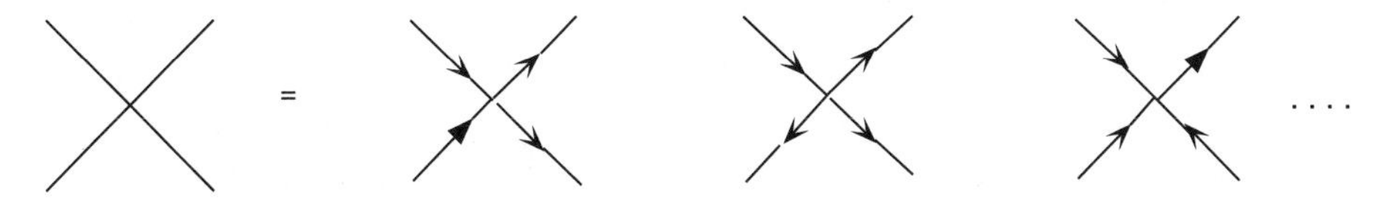

Figure 9.1 The basic "four-meson" interaction of the model theory (left diagram). It subsumes five different fundamental transformations, three of which are shown on the right.

Feynman Diagrams

Field theory got its start in the late 1920s with quantum electrodynamics (QED). The early treatment of QED involved a mixture of quantum particle mechanics for electrons and other charged, material particles, and quantum field theory for the electromagnetic field with photons arising as field quanta. In time, quantum fields were introduced for other particles as well; and we now believe that all particles are the quanta of fields. Perturbation theory was developed in concert with QED. It was applied to a growing body of experimental information on various electrodynamic processes; for example, photon-electron and electron-electron scattering reactions. The mathematical technology of perturbation theory was well-defined and unambiguous when carried out to lowest relevant order (lowest relevant power of the coupling constant); and already in lowest order theory and experiment agreed quite well. This seemed reasonable inasmuch as the expansion parameter for electrodynamic cross sections is $e^2/\hbar c \approx 1/137$, a small number. But in higher orders the computations produced infinite cofactors, a problem that seemed to be characteristic of quantum field theories in general. For some people, including many of the founders of the subject, this suggested that the concepts of quantum field theory were in need of fundamental revision. But those concepts were given new life as a new generation of practitioners learned how to isolate the infinities and absorb them into a few basic parameters of the theory in a procedure known as "renormalization." This may have looked to some to be a sweeping of infinities under the rug, but it worked spectacularly well, producing some of the most accurate items of agreement between theory and experiment known to science. For renormalizable theories such as QED, the mathematical procedures are complicated but unambiguous to all orders. They took their final shape during a hectic period around the end of the 1940s. The most convenient and picturesque formulation was developed by a famously colorful young physicist, Richard Feynman. Julian Schwinger, who independently established the theory in an equivalent but more complex formulation, described Feynman as "bringing computation to the masses."

The transition amplitude—the Feynman amplitude as it is often called—for any particular process is the sum of an infinite number of terms. Each term can be visualized by means of a Feynman diagram, which lends itself to a rough but intuitive physical interpretation. Diagrams are composed of vertices and lines. In the model theory we've been discussing there are four lines joined together at each vertex. The lines either connect to other vertices or hang freely, each free line representing one of the particles involved in the process under consideration. A line that connects one vertex to another is called a *propagator*. It describes the propagation of a virtual meson that is created at one space-time point and destroyed at another. In the model theory the contribution to the Feynman amplitude coming from a diagram containing n vertices is proportional to λ^n, the nth power of the coupling constant. At each vertex, the interaction Hamiltonian does its work, creating and/or destroying real or virtual mesons in the manner described above. To illustrate, consider a reaction in which two mesons scatter elastically. We denote this by $a+b \rightarrow c+d$. There is only one species of meson in this model, so the labels here are not meant to distinguish among species. Rather, they are shorthand for the momenta of the mesons: the letters a, b, c, d stand respectively for the momenta $\mathbf{p}_a$ and $\mathbf{p}_b$ of the incident mesons, $\mathbf{p}_c$, and $\mathbf{p}_d$ the momenta of the outgoing mesons. Of course, momentum conservation dictates that $\mathbf{p}_a + \mathbf{p}_b = \mathbf{p}_c + \mathbf{p}_d$, and energy conservation that $E_a + E_b = E_c + E_d$.

To first order in the coupling constant λ, there is only the $2 \rightarrow 2$ diagram of Fig. 9.1, the incoming legs being labeled by the letters a, b and the outgoing by c, d. The reaction $a+b \rightarrow c+d$ proceeds directly here. The interaction Hamiltonian acts once, destroying the incident mesons and creating the outgoing ones. The Feynman amplitude corresponding to this diagram is as simple as can be. It is just equal to λ, independent of energy and scattering angle. The drawing labeled α on the left in Fig. 9.2 is a second-order diagram. It has two vertices and so contributes

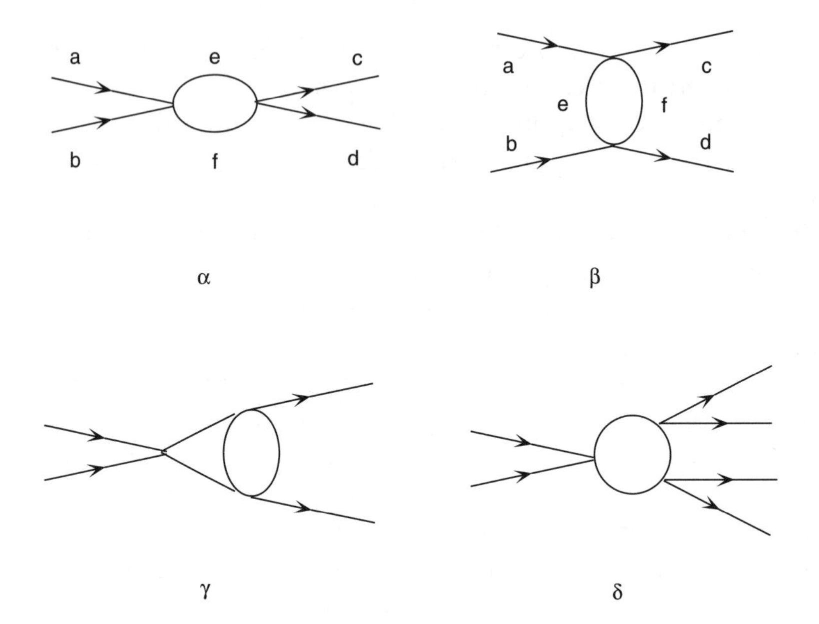

Figure 9.2 α and β: Second-order Feynman diagrams for elastic scattering in the model theory; γ: One of several third-order diagrams. δ: One of several lowest-order diagrams for the two-meson to four-meson reaction.

to the amplitude a term proportional to λ^2, now multiplied by a function of the collision energy and scattering angle. The rules for deducing this latter function are technical and complicated, but there is a simple physical interpretation that goes with the drawing. The labels e and f attached to the "internal" lines represent virtual mesons. They propagate from one vertex to the other. The one diagram α actually subsumes two different sequences: (i) The incident mesons a and b collide at the left-hand vertex to produce a pair of virtual mesons e and f; the latter then propagate to the space-time point represented by the right-hand vertex and collide there to produce the outgoing mesons c and d. We symbolize this by the sequence $a+b \to e+f$ followed by $e+f \to c+d$. Each step in the sequence is a $2 \to 2$ transition; that is, two mesons are destroyed, two are created. If arrows were to be supplied for the virtual mesons, the arrows would point from left to right in the diagram. (ii) The other sequence corresponds to the interaction Hamiltonian creating, out of nothing, the four mesons c, d, e, f, the incident mesons a and b remaining untouched at this stage; this followed by the destruction of a, b, e, f. The two steps in this sequence consist therefore of the $0 \to 4$ transition $0 \to c + d + e + f$ followed by the $4 \to 0$ transition $a + b + e + f \to 0$, where the symbol 0 stands for "nothing." The above two sequences are subsumed, as was said, in the one diagram α of Fig. 9.2. The drawing β in Fig. 9.2 is another second-order diagram. It too subsumes a pair of sequences: (i) In a $1 \to 3$ transition, the incoming meson a dissociates into the outgoing meson c and the virtual pair $e+f$; then e and f meet up with b and all three are destroyed in the $3 \to 1$ transition creating d. We symbolize this sequence by $a \to c+e+f$ followed by $b+e+f \to d$. (ii) The other sequence corresponds to $b \to d + e + f$ followed by $a + e + f \to c$.

The drawing γ in Fig. 9.2 is one of several *third*-order diagrams (it has three vertices). We will mercifully refrain from describing the sequences that it subsumes, but that should not be a difficult task for the committed reader. Finally, the drawing δ in Fig. 9.2 is one of several lowest-order diagrams (three vertices, hence third order) for the multiparticle production reaction $a + b \to c + d + g + h$ in which two mesons collide to produce four mesons. The reader is invited to construct one or two other diagrams.

Virtual Particles

Experts in the subject will know, just by looking at any Feynman diagram, what computations have to be performed, though they may shudder at the prospect if the diagram is complicated. Any given propagator has a prescribed dependence on the energy and momentum variables of the virtual particle that is propagating from one vertex to another. Generally, the computation involves integration over these variables. The higher the order, the more diagrams there are and also the more variables to be integrated over. But this drudgery aside, for our purposes the main insights are qualitative, a recognition of the sequences of elementary transitions that compound to produce some physical reaction. The virtual particle concept that is involved in all of this is quite fascinating. The "real" particles in a given reaction are the incident ones that were prepared far apart from one another and then brought into collision; and the outgoing ones, detected when they have moved far apart from one another. In the course of the collision, when everything is in close contact, virtual particles come and go. They are intermediaries in any given physical reaction. There are two different ways of describing their stance with respect to energy conservation. In the language that was adopted above, the language in which a given Feynman diagram is said to correspond to several different sequences of elementary transitions, energy conservation (though not momentum conservation) is violated at any vertex involving at least one virtual particle. This is no cause for alarm, however. Even if the "real" particles are stable, in their virtual incarnations they have only a transitory existence. A virtual particle that comes into existence for a time interval Δt must necessarily have an energy spread ΔE no smaller than that given by the "uncertainty relation" $\Delta E \Delta t \approx \hbar$.

However, there is another way of organizing the calculations. It turns out to be mathematically rewarding to group together the contributions from certain elementary transitions. In this way of proceeding, which was developed especially by Feynman, energy as well as momentum is conserved at all vertices. But now the virtual particles no longer have a definite mass. Rather, for each virtual particle the mass becomes, in effect, one of the integration variables. Thus, in the one way of assembling the calculations the virtual particles have the correct mass but they violate energy conservation. In the other way, energy and momentum are strictly conserved at each vertex but the mass of a virtual particle is a variable. There is no contradiction in the final outcome between these two ways of looking at things. They simply correspond to two different ways of arranging the calculation of a Feynman amplitude. The energy-violating procedure lends itself more readily to physical interpretation, the Feynman approach more readily to efficient calculation. What this brings out, however, is that the virtual particle concept is actually only a proxy for certain mathematical ingredients, though an intuitively helpful proxy; and that different ways of organizing the mathematics correspond to different versions of the proxy. Virtual particles, after all, are not real objects. It might be a good compromise to describe them as corresponding to virtual reality.

It was said above that virtual particles come into play when the real collision ingredients are all close together. In fact, virtual particles are always in play. Even a single real particle, moving along in isolation, can emit and reabsorb virtual particles, over and over and over again. This has the effect of shifting the physical mass of the particle away from the "bare" value that entered into the Hamiltonian. That shift inevitably turns out to be slightly infinite, and there is a whole technology for isolating and redefining away this and a few other infinities that are characteristic of renormalizable quantum field theories. But we forebear to pursue these delicacies any further here.

The Standard Model in Diagrams

The Basic Interactions

The model field theory that we have been describing for illustrative purposes is not at all realistic. But as was said earlier, it is easy enough at the free field level to set up a theory based on fields we believe to be more realistic: fields corresponding to the quarks, leptons, gauge bosons, Higgs boson, and perhaps other particles that will be prompted by new experimental discoveries or compelling theoretical ideas. At the free field level, however, nothing happens. Happenings are induced by interactions among the fields, terms that couple the fields together in the Hamiltonian. These constitute the field theoretic analog of forces in particle mechanics. In the last chapter we have already described in words some of the basic interactions incorporated in the modern theory. We will be repeating many of these words here; but the basic interactions can now also be displayed diagrammatically, as in Fig. 9.3. This is not quite the full set of interactions, but enough is shown to indicate the main features. Quarks, gluons, and charged leptons are denoted by the letters q, g, and ℓ; the neutrino associated with charged lepton of type ℓ, by ν_ℓ; the photon and charged and neutral weak interaction bosons by the letters γ, W, and Z. The words and symbols are used here in a collective sense to include particle and antiparticle wherever they are distinct.

The "strong" diagrams in Fig. 9.3 represent the basic interactions of quantum chromodynamics. The top figure depicts the coupling of a pair of quarks to a gluon; the other figures depict interactions solely among the gluons. Notice in particular that the top diagram subsumes a collection of fundamental processes. For each of the six quark flavors, $q \leftrightarrow q + g$, $\bar{q} \leftrightarrow \bar{q} + g$, $q + \bar{q} \leftrightarrow g$, $q + \bar{q} + g \leftrightarrow 0$. In this context q stands for a quark particle, $\bar{q}$ for its antiparticle. The coupling constant is independent of the quark flavor. Indeed, all the strong interactions of Fig. 9.3 are parameterized by a single, strong interaction coupling constant.

The single "electromagnetic" diagram of Fig. 9.3 represents the basic interaction of a charged particle with the photon. *Any* electrically charged object Q, simply by virtue of its charge, necessarily couples to the photon. The magnitude of the coupling constant is just the electric charge. For all the fundamental particles we deal with, this magnitude is just the charge e on a proton, or, for quarks, a not-too-small fraction thereof, 2/3 or 1/3. The diagram in Fig. 9.3 subsumes the transitions $Q^{\pm} \leftrightarrow Q^{\pm} + \gamma$, $Q^{+} + Q^{-} \leftrightarrow \gamma$, $Q^{+} + Q^{-} + \gamma \leftrightarrow 0$.

The remaining drawings in Fig. 9.3 depict the weak interactions. The diagram on the lower left represents the coupling of quarks to the charged vector bosons W:

$$(u, c, t) \leftrightarrow (d, s, b) + W^{+}, \quad (\bar{u}, \bar{c}, \bar{t}) \leftrightarrow (\bar{d}, \bar{s}, \bar{b}) + W^{-},$$

$$(u, c, t) + (\bar{d}, \bar{s}, \bar{b}) \leftrightarrow W^{+},$$

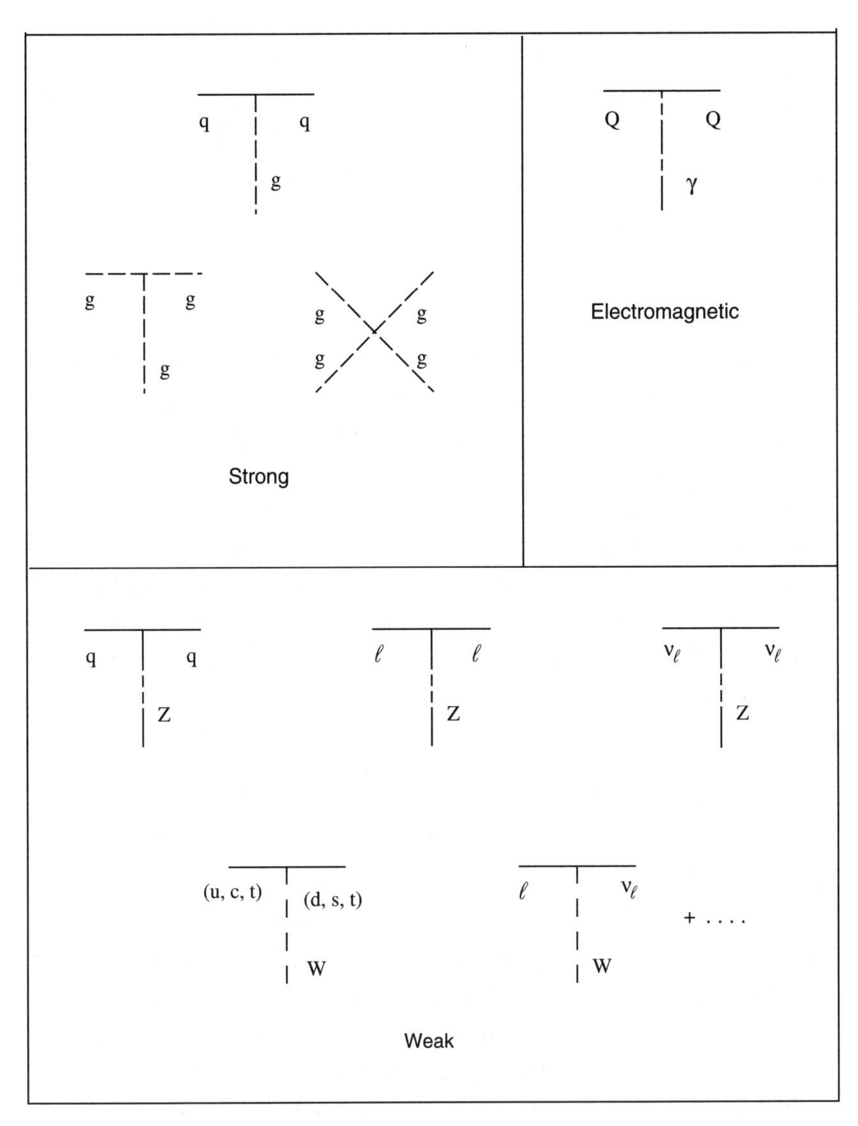

Figure 9.3 Some fundamental interactions of contemporary particle theory. The symbols q, ℓ, ν_ℓ, denote quarks, charged leptons, neutrinos, and their antiparticles. The symbols g, γ, Z, W denote gluons, photon, neutral weak boson Z, and charged weak bosons W^+ and W^-. The symbol Q represents any charged particle.

and so on (by now the meaning of "and so on" should be clear). The notation here is meant to indicate that the u quark, for example, can transform to any of d, s, b; so too for the c and t quarks. Mainly, however, u prefers to go to d, c to s, and t to b. The diagram on the lower right depicts the coupling of a charged lepton and its neutrino to the vector bosons W:

$$\ell^- \leftrightarrow W^- + \nu_\ell, \quad \ell^+ \leftrightarrow W^+ + \bar{\nu}_\ell,$$

$$W^- \leftrightarrow \ell^- + \bar{\nu}_\ell, \quad W^+ \leftrightarrow \ell^+ + \nu_\ell,$$

and so on, where $\ell = e, \mu, \tau$. The other "weak" diagrams correspond, respectively, to the coupling of quarks, charged leptons, and neutrinos to the neutral Z boson: $q \leftrightarrow q + Z$, $\ell \leftrightarrow \ell + Z$, $\nu_\ell \leftrightarrow \nu_\ell + Z$, and so on.

The basic weak interactions outlined above are all controlled by coupling constants that have roughly the same magnitude as the characteristic electromagnetic constant; namely, the charge on a proton. As noted earlier, this reflects a deeper aspect of the modern theory, the unification of weak and electromagnetic interactions.

Collision and Decay Reactions

The basic interactions outlined above constitute a kit of parts with which to construct various reaction processes. For an immediate example consider electron-positron annihilation into a pair of oppositely charged muons: $e^- + e^+ \rightarrow \mu^- + \mu^+$. There are of course infinitely many Feynman diagrams for this reaction as for any other, but since the controlling coupling constant here is small, it is a good approximation to restrict ourselves to the lowest-order Feynman diagram on the left in Fig. 9.4. It

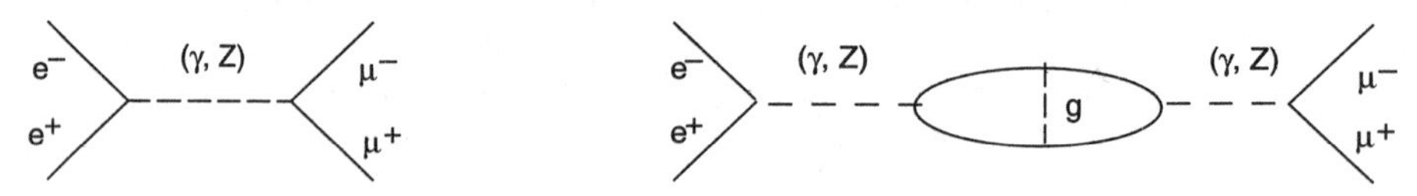

Figure 9.4 Feynman diagrams for the process $e^- + e^+ \rightarrow \mu^- + \mu^+$.

really summarizes two different diagrams. One involves a virtual photon (or, as one sometimes says, an *intermediate* photon); the other, an intermediate Z boson. Each of these has two vertices, so the corresponding amplitudes are both proportional to e^2. The photon and Z boson propagators differ only because the photon and Z boson masses are so different. For either, the propagator is of the form propagator = $[(\text{energy})^2 - (\text{mass})^2]^{-1}$, where "energy" refers to the total center of mass energy W of the collision and "mass" refers to the mass of the intermediate particle. The photon mass is of course zero, whereas the Z boson mass M is very large. Up to a roughly common proportionality factor, therefore, the two amplitudes are

$$\text{amp}(\gamma) \approx e^2/W^2, \quad \text{amp}(Z) \approx e^2/(W^2 - M^2).$$

Although the weak and electromagnetic interactions have roughly the same coupling constant e, it is evident that at low energies, $W \ll M$, the weak amplitude (the one involving the Z boson) is selectively suppressed. At very high energies, $W \gg M$, the two amplitudes are comparable. It should be confessed that our formula for the Z boson propagator has been simplified a bit. It really doesn't become infinite when $W = M$, though it does become large at and near that energy.

What the above example illustrates is a rather general feature of weak versus electromagnetic interactions. The basic coupling constants are comparable, but for low-energy processes the weak amplitudes are suppressed because large W or Z boson masses inevitably appear in denominators of propagators. We may also use the process $e^- + e^+ \rightarrow \mu^- + \mu^+$ to bring out another point; namely, that all three classes of basic interactions, strong, electromagnetic, and weak, inevitably enter into all possible reactions. In the left-hand diagram of Fig. 9.4 a gauge boson (photon or Z boson) is exchanged between the electron-positron pair and muon-antimuon pair. In the diagram on the right side of Fig. 9.4 the intermediate gauge particle decides en route to convert to a *quark-antiquark* pair which then annihilates to restore a gauge boson. But the quark and antiquark along their way decide to exchange a gluon. Strong vertices have hereby entered the picture. Even so, the amplitude coming from the right-hand diagram makes only a small contribution. That's because it is proportional to the fourth (rather than second) power of the small electroweak coupling constant e (four of the vertices are electroweak).

Given the kit of basic vertices, it is easy enough to draw the lowest-order diagrams for any collision or decay reaction among quarks, leptons, and gauge bosons. A few additional examples are collected in Fig. 9.5. One of the diagrams depicts the decay reaction $\mu^- \rightarrow e^- + \nu_e + \nu_\mu$ in lowest order; another is one of several lowest-order diagrams for $e^- + e^+ \rightarrow 3\gamma$; the third is the lowest-order diagram for the scattering of light by light, $\gamma + \gamma \rightarrow \gamma + \gamma$. It is also easy enough to draw higher-order diagrams for these or any other processes, although the number of diagrams grows rapidly with order.

The perturbation theory approach embodied in Feynman diagrams has a major limitation. The hadrons—proton, neutron, pi mesons, and all the others—do not show up in those diagrams. That's because the hadrons are not on our list of elementary particles. They are bound states of quarks and gluons, and the perturbation approach is not of much use. For example, the internal wave function of a π^+ meson necessarily has a $u\bar{d}$ component, but it also contains admixtures of various numbers of gluons and same-flavor quark-antiquark pairs, mainly $u\bar{u}$, $d\bar{d}$, $s\bar{s}$, and so on. A great deal of empirical information has been acquired about the internal structure of protons and neutrons, but a purely theoretical determination is no easy matter, although progress is being made. The Feynman diagram approach, then, is of limited quantitative usefulness for

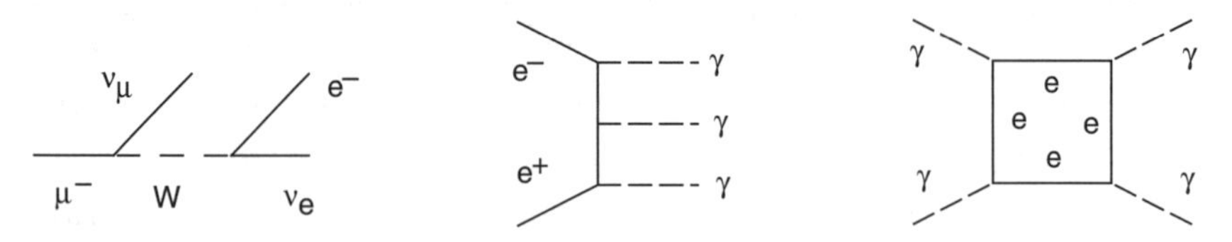

Figure 9.5 Lowest-order Feynman diagram for several different processes.

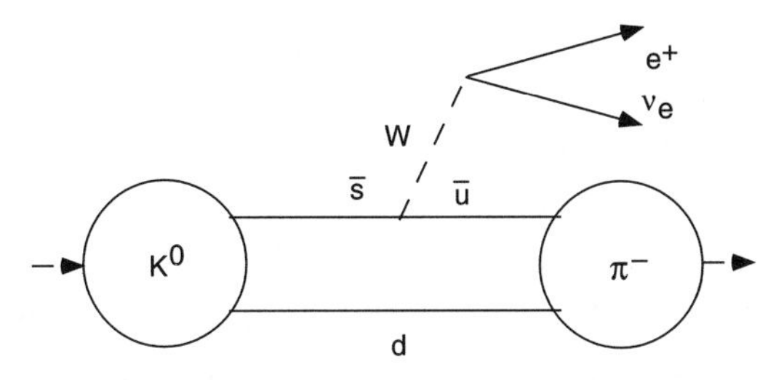

Figure 9.6 Feynman diagram depicting $K^0 \to \pi^- + e^+ + \nu_e$, with black boxes for the hadrons.

reactions involving hadrons. But the diagrams can still serve qualitatively. A single example will suffice. Consider the weak decay reaction $K^0 \to \pi^- + e^+ + \nu_e$. Take the neutral K meson to be predominantly $d\bar{s}$, the negative pion, $d\bar{u}$. The reaction can then be pictured as in Fig. 9.6. There is a single weak vertex here—it is certainly justified to work to lowest order in the electroweak couplings. But strong interactions are occurring all over the place within those black boxes representing the K^0 and π^- mesons.

Again, What's Going On?

Quantum theory copes with a number of miracles, many of them outrageous to common sense and intuition, others so familiar that one can easily fail to notice and marvel. The ancient question whether matter is continuously divisible or built up of discrete, fundamental entities was resolved decisively by the early years of the twentieth century in favor of the atomic hypothesis. To be sure, chemical atoms did not turn out to be the ultimate building blocks, nor did the proton and neutron ingredients of the chemical atom. Rather, the onion has now been peeled down to the quarks, leptons, and gauge bosons that we have been discussing. There are reasons to think that these entries on the list of fundamental particles may yet be added to. There is very little indication at present that any of the entries can themselves be peeled down any further, though that is certainly not inconceivable. In any case, the main point is that the onion is not *continuously* layered; the layers are discrete.

Already in the nineteenth century a few perceptive observers were struck not just by the growing success of the atomic picture but by the miracle that all the atoms of a given species seem to be identical. Whatever the ingredients of an atom or molecule might prove to be, from a classical point of view one would expect a continuous range of possible internal configurations and hence binding energies, chemical properties, and so on. Why, for a given chemical species, should the ingredients have assembled into the same internal configurations for all the world's atoms of a given species? In his *Encyclopedia Britannica* article on atoms and molecules, the great James Clerk Maxwell wrote that "the formation of a molecule is therefore an event not belonging to the order of nature in which we live" but must instead be referred back to an epoch of "the establishment of the existing order of nature...." We have seen how this conundrum is resolved in quantum mechanics. Instead of a continuum of possible configurations of the ingredients of an atom or molecule there is a discrete spectrum of bound quantum states. If two molecules are in different states they are in fact *not* identical; but if they are in the same quantum state, they *are* identical. That is, what is identical among all the members of a given species is the spectrum of states. All of this is just an incidental consequence of the remarkable fact of quantization of bound states. But there is a deeper miracle; namely, that all the members of a given species of atom are made of identical ingredients—thus, that all the electrons of the world, all the protons, all the neutrons are identical; and, deeper down, that all the quarks of a given flavor and color and all the gluons of a given color class are identical.

This property of identity among the building blocks of a given species can simply be postulated, as is done at the level of quantum particle mechanics. But it emerges automatically from the application of quantum principles to fields. This is one of the great, often unsung triumphs of quantum field theory. At the classical level, particles and fields coexist on equal terms. At the quantum level, it is the fields that are primary. Particles emerge as the quanta of fields, and in identical copies.

Most remarkable of all is that matter can be created and destroyed—not merely rearranged, but created and destroyed. Quantum field theory appears to provide an adequate theoretical framework for dealing with this. Popular accounts of quantum mechanics will often incorporate this fact well enough, but they do not so often marvel at it sufficiently. The ancient idea of irreducible building blocks of the material world has vanished! Those same popular accounts will often describe the creation process, with much appeal to $E = mc^2$, as the conversion of energy into matter; conversely, the destruction process is pictured as the transformation from matter into energy. A favorite example is the dread annihilation of matter and antimatter. But this is quite misleading. It is true that particle reactions, indeed all transformations generally, must respect energy and other conservation laws. But energy is not some disembodied effluvium. It is borne in the rest and motion energies of the real, physical particles that participate in a reaction. Thus, when a proton and antiproton "annihilate," real things emerge; for example, pions, as in the reaction $p + \bar{p} \rightarrow \pi^+ + \pi^-$. The total energy is the same on both sides of the equation. Annihilation reactions are no different than other reactions in which particles are created and destroyed. Indeed, even if a certain incident particle reemerges in the outgoing state of a reaction, it is best to think of it as having been destroyed and recreated in the process.

Creation and destruction of matter is an awesome thing. Quantum field theory provides what seems to be the appropriate conceptual and mathematical machinery, although we could only barely skim the subject: classical fields converted into quantum field *operators*; interaction terms in the Hamiltonian playing the role of forces and acting on particle-like energy-momentum eigenstates to produce new states with altered particle content; and so on. The trouble is, this all seems so unphysical, formal, bloodless. What is *really* going on? Feynman's diagrams help a little. At each vertex real and virtual particles are created or destroyed, the virtual particles travel to other space-time points where this is repeated, and over and over. Any physical reaction is the sum of the different pathways charted by the infinitely many Feynman diagrams. But of course this doesn't "explain" how those basic acts of creation and destruction at the individual vertices occur in the first place. Going back to a classical picture, one might speculate as follows. Maybe there are no material particles at all, only fields. Maybe what we think of as particles are really only regions of concentrated field strength. It is easy enough in classical field theory to imagine that localized disturbances can fragment into other localized disturbances, or collide and change form and multiplicity, and so on. Think of waves crashing on waves in a stormy sea. But these are only idle, armchair musings. Nothing remotely realistic has ever gotten anywhere along these lines.

The best explanation of creation-destruction and all the other wonders of the quantum world may well be Feynman's. To paraphrase him: that's the way the world is.

READINGS

A Small, Personal Selection

Pais, A. *Subtle Is the Lord*. Oxford University Press, 1982. This is the classic scientific biography of Albert Einstein; a superb source of information and insights on the origins of quantum mechanics in the blackbody radiation problem and on Einstein's lifelong struggles with the quantum.

Cline, B. *Men Who Made a New Physics*. University of Chicago Press, 1987. A not-too-technical history of the foundation, the founders, and their interpretative views.

Jammer, M. *The Conceptual Development of Quantum Mechanics*. Wiley, 1974. A scholarly and authoritative account, in prose as well as equations.

Schweber, S. *QED and the Men Who Made It*. Princeton University Press, 1994. A history of quantum field theory, much of it technical but interspersed with fascinating sketches of the founders and other leading personalities.

Wheeler, J. A., and W. H. Zurek, eds. *Quantum Theory and Measurement*. Princeton University Press, 1983. A great collection of classic papers, many by the masters, on the conundra and problems of interpretation of quantum mechanics.

Hey, T., and P. Walter. *The Quantum Universe*. Cambridge University Press, 1987. The authors provide a delightful description of the structure, applications, and wonders of quantum mechanics. Witty, not too technical, marvelous photographs and cartoons.

Feynman, R. *QED, The Strange Theory of Light and Matter*. Princeton University Press, 1985. Quantum electrodynamics explained in layman's terms by the master.

Pagels, H. R. *The Cosmic Code: Quantum Physics as the Image of Nature*. Simon and Schuster, 1982. A nonmathematical description of the quantum world and its conundra.

Zee, A. *Fearful Symmetry*. Macmillan, 1986. A chatty, deeply insightful paeon to symmetry as a guide to discovery of Nature's laws.

Wilczek, F., and B. Devine. *Longing for the Harmonies: Themes and Variations from Modern Physics*. Norton, 1988. A delightful, authoritative, and highly original collection of short pieces ranging over the wide landscape.

Weinberg, S. *The Discovery of Subatomic Particles*. Freeman, 1983. The growth of the atomic hypothesis; discoveries of the electron, nuclear atom, neutron; and more. Accessible and beautifully well told.

Bernstein, J. *The Tenth Dimension*. McGraw Hill, 1989. An informal, fairly detailed account of particle physics, told in prose.

Ne'eman, Y., and Y. Kirsh. *The Particle Hunters*. Cambridge University Press, 1996. This broad survey ranges from the early atoms on through the modern standard model and beyond.

INDEX